SUSTAINABLE
SUCCESS

SUSTAINABLE
SUCCESS

How Businesses Win as a Force for Good

Paul Marushka

Rivertowns
BOOKS
IRVINGTON, NEW YORK

Printed in the United States of America · April 2025 · I

Hardcover edition ISBN-13: 978-1-953943-58-3
Paperback edition ISBN-13: 978-1-953943-59-0
Ebook edition ISBN-13: 978-1-953943-60-6

LCCN Imprint Name: Rivertowns Books
Library of Congress Control Number: 2025930816

Rivertowns Books are available from all bookshops, other stores that carry books, online retailers, and directly from the publisher. Visit our website at www.rivertownsbooks.com. Retailer and consumer orders, inquiries about discounts for bulk purchases, and other correspondence may be addressed to:

Rivertowns Books
240 Locust Lane
Irvington NY 10533
Email: info@rivertownsbooks.com

To Emilia, Sonya, Luke, and Alexander with all my love and thanks.
You inspire my own efforts to be a Force for Good.

CONTENTS

FOREWORD

Dane Parker

Former Chief Sustainability Officer at General Motors and Board Director at Sphera, Bridgestone Americas, and Lion Electric

I was honored when Paul asked me to write the foreword to his book. I have a huge amount of respect for what he's doing at Sphera—using data to help companies create a safer, more sustainable and productive world. Paul is a people-focused leader and is culturally aware. He balances competing objectives wisely. Importantly, for the reader, Paul isn't somebody who sits in a particular camp; his work in data cuts across all sectors. He's not running for office or managing a non-governmental organization (NGO) with a particular point of view. This neutrality gives him a position of credibility when it comes to positively influencing others.

Therefore, his book provides a view of the business world that is rational and measured, driven by facts rather than emotion. This is a depolarized zone.

Theory for Change

In these pages, Paul takes the reader on a journey through industrial history, showing how cultural shifts have constantly improved the world. Time and time again, it is business that makes those changes real and permanent. The first movers and fastest followers are the ones that recognize the shifting sands, becoming a Force for Good in society. Those businesses that can't or won't react are often consigned to the footnotes of history.

Paul describes in the book how the definition of *good* has changed through history—and it will change again. Personally, I hope the next definition will be focused on humanity-based sustainability: a more inclusive society, better living conditions, access to food, clean water, affordable energy, and lifting people out of poverty.

There is so much in this book to provoke thought among CEOs on the front lines of business today. Alongside the stories and personalities, there are plenty of how-to takeaways for discussion in the next board meeting. The interactive elements of the book allow the reader to diagnose their own sustainability condition. I find the Getting Started, Getting By, and Getting Ahead barometer in chapter six especially insightful and useful.

I believe that good CEOs are the ones who think about their legacy. They ask themselves, "What are people going to say about me when I leave?"

Some will want to be known for making a lot of money for the company. Fine. That's one strategy, although there are countless examples of how short-termism focused only on financial results has set a company on course for a train wreck.

But if you want people to say you were a Force for Good, then you're going to have to consider an alternative strategy, including actions that are more aligned with long-term company sustainability.

I've been fortunate to observe and work for some excellent CEOs who played the long game and saw the bigger picture. Andy Grove at Intel, where I started my career, had an innate willingness to change. He created a culture to encourage innovation, experimentation, and a forward-looking perspective.

At GM, Mary Barra was the first female CEO of a major auto company. Her drive to innovate, which has proved critical over the last decade, stems from her desire to create a company that will last another 100 years. The women I've worked with have always been more open to other ideas than their male counterparts. Dare I say it, they have less of an ego. I think it would be marvelous for industry to have more senior leaders who are women.

Searching for Balance

I know firsthand from my time at GM that many large manufacturers want to be successful as much for their communities and employees as for their shareholders. That's a challenge, because sometimes the long-term needs of the former clash with the short-term demands of the latter. But good companies find the right balance.

Increasingly, the needs of the environment and society overlap with those of customers and investors. Regulations will only push a company so far. The tipping point is when customers and investors turn off the cash. Then CEOs take notice. At the same time, consumers, investors—and governments, too—will reward companies that do the right things.

CEOs need an innate willingness to change. A key to avoiding complacency bias is to have a clear purpose—what Paul describes in the book as a Noble Purpose—and the inclination to listen. I have always liked this observation by Mahatma Gandhi: "I do not want my house to be walled in on all sides and my windows to be stuffed. I want the culture of all lands to be blown about my house as freely as possible. But I refuse to be blown off my feet by any."

Business leaders should bask in the winds of change. The best will recognize the need to adapt, embrace new technologies, and bring in fresh talent. But they will also stay true to the core principles of the business. With a strong base, you can adjust.

As a board member for the Americas division of global tire manufacturer Bridgestone, I've gained a fascinating insight into an enterprise that is guided by a Noble Purpose. Shojiro Ishibashi founded the business in Japan in 1931 with a clear business philosophy: "I am convinced that a simple profit-seeking business will never thrive, but a business that contributes to its society and country will be forever profitable." That visionary ethos holds true to this day and guides the company's strategy and decisions. What CEO doesn't want to run a business that becomes forever profitable?

Another business leader I look up to is Jim Goodnight, cofounder and CEO of the SAS Institute since 1976. The statistical analysis software company has increased revenues and turned a profit for 40 years, but Goodnight has repeatedly put his own values ahead of breakneck growth. Competitors have come and gone, while SAS has continued to adapt and innovate to remain a respected leader in its field. If you're looking for a leader with a great legacy, then look no further.

No Company Is an Island

In this book, Paul talks about the need for businesses to get ahead of cultural change in pursuit of the next big innovation that's going to stick. I completely agree with this sentiment. Thinking through a cultural lens is so important for CEOs, perhaps more so today than ever before. Challenges with the workforce and the explosion of information in the media are driving and influencing culture all the time. If you leave those out of your company narrative, then you are left outside in the cold.

The hard-nosed, my-way-or-the-highway bosses are mainstays of certain Netflix series, but they are an endangered species in today's boardrooms. They are missing out on the opportunities offered by diversity, social change, and the power of unlocking almost limitless human potential.

Paul's optimism about the future and his belief in the power of innovation are infectious. The book has made me ponder what the world is going to look like in 2050. I feel reassured by the resilience—even the *antifragility,* to borrow a term from Nassim Nicholas Taleb—of the human race.

People increasingly understand that no country or company is an island. The conversation about the environment will (I hope) turn toward possibility, and some of the fear will subside. I'm confident that tomorrow's behaviors across the whole world will create a more sustainable path forward, from the way we shop and eat to where we live, work, and travel.

We'll see less poverty and greater levels of industrialization in places around the world that are now far behind. Those developing countries will almost certainly progress in a more future-focused way than we did in the West the first time around.

Those leading companies of today that survive and thrive into the second half of the 21st century will keep on adapting. They will move into new industries. They will become competitive in data and digitization because there's so much profitability in those businesses. But I hope, at the same time, they won't give up on making actual products that people need and want. Manufacturing provides meaningful work opportunities for people with a variety of skills and levels of education and is often a model of diversity in the workforce. If there's one thorn that could burst my 2050 balloon, it would be a failure to build a sustainable workforce. But that's the content of a whole other book.

I recently had a discussion about business and politics with the CEO of a major company. We lamented the current leadership vacuum in society. Politics are becoming increasingly distasteful. Businesses must take the opportunity to step up. The public today has more trust in business than elected officials or the media, so CEOs need to repay and consolidate that trust.

Information companies like Sphera can play their part, as they provide an agnostic lens on the world. They deal in truth alone. Effective data is critical for measuring performance across sustainability considerations, which are now so important for company reputation and operational productivity.

Data gives you a view of where you are and where you're going (the why needs to come from inside). It gives you the good, the bad, the ugly, and the wonderful. If you know exactly where you're standing, you can articulate your position and defend your reality as you express a vision for the future.

I strongly believe that businesses have an opportunity to lead in shaping the future. Businesses hire people, they pay people, they provide jobs, they help families make a living. They manufacture the things we buy and need. They make life better. Businesses hold a position of importance and value in people's lives. This privilege presents an opportunity—perhaps even an obligation. Shame on us if we can't tackle the challenges of our times.

Paul is right. Being a Force for Good is good for business. I'd like to add that business must be a force for *the better* too. Because the prospects for humanity-focused sustainability are vast.

ACKNOWLEDGMENTS

I'm sure anybody who has ever written a book will tell you that it's no small undertaking. It's a long journey, with many twists and turns, at the end of which you get to hold something in your hands of which you can be really proud.

I want to take a moment to draw readers' attention to some of the many people without whose contributions I am certain this work would be much diminished. They include:

- Lisa Agona
- Mikaela Algren
- Kris Dubey
- Megan Geldman
- Constantine Limberakis
- VeeAnder Mealing
- Andy North
- Alex Studd
- Jenn Tabba
- Madeline Temple
- Mike Zamis

For all of their insights, ideas, and inspiration, as well as the support they provided in the process of creating and publishing this book, I thank them.

I also want to thank the contributors whose biographies appear toward the end of the book. They are a remarkable collection of individuals who are thought leaders in many aspects of the sustainable journey that organizations today must take. Each one gave generously of their time as I researched this book, sharing some of their most valuable ideas and experiences for the benefit of readers. I'm very grateful for their help.

Paul Marushka

THE INVENTION TEST: PART 1

H umanity's desire to improve the world we live in has led to some re-markable technological advances. As we will discuss in detail in the coming chapters, business leaders have often designed these inventions in response to the cultural and social needs of the day, driven by a Noble Purpose. Those who couldn't or wouldn't adapt have earned a more notorious reputa-tion as laggards on the wrong side of history.

But before we delve into these stories and the insights they offer for busi-ness leaders today, let's take a quick quiz. Who doesn't enjoy a little trivia con-test to get the synapses firing? And this one is offered just for fun—there are no prizes, and no forfeits either.

We've come up with a collection of multiple-choice questions that will test your knowledge of famous high achievers from the world of invention. Every few chapters, you'll be challenged with half a dozen questions as you make your way through the book. Along the way, you'll discover the Lions who roared into the annals of innovation—and a few unfortunate Ostriches who sank their heads in the sand.

If you enjoy this challenge, you may want to invite your friends and col-leagues to test their own prowess. They can access our digital version of the test at https://sustainablesuccess.sphera.com/theinventiontest—or, of course, they can buy their own copy of this book.

The Invention Test gives you a taste of the kinds of characters and stories that we'll encounter in the pages ahead. We'll explore why some business leaders can surf the waves of a changing culture, while others are thrown onto the beach of broken dreams. What's the formula for success when it comes to mastering innovation? And how should CEOs be setting their strategic priori-ties today and into the future? Those are the questions we'll tackle in the chapters that follow.

Question 1

Which company's purpose reads, "We're in business to save our home planet"?

A. Tesla
B. Patagonia

The answer is B: Patagonia

For over 50 years, the outdoor apparel company Patagonia has sought to hold true to its Noble Purpose of changing the way business is done for the benefit of the environment and ecology. In 2022, founder Yvon Chouinard, himself a passionate rock climber and surfer, took the unprecedented step of bequeathing the whole business to an environmental trust, making Earth the "only shareholder." The company remains for profit, but with the aim of "proving that purpose and profits are inextricably linked."

Tesla's purpose is "to accelerate the world's transition to sustainable energy" through its electric vehicles and renewable energy generation and storage. In 2014, the company chose to open-source its battery patents with the aim of fast-tracking their mass adoption. The gesture has helped scale the battery energy storage industry globally. The cost of one kilowatt-hour (kWh) of lithium-ion battery capacity fell from $1,200 in 2010 to under $140 in 2023. The price is predicted to drop further to US$62/kWh by 2035.

Question 2

What "cool" invention did Willis Carrier create in 1902?

A. The first refrigerator
B. The first air-conditioning system

The answer is B: the first air-conditioning system

O n a chilly autumn evening in 1902, New Yorker Willis Carrier was waiting for a train home from Pittsburgh on a platform filled with mist from a surrounding fog bank. Like other ambitious engineers of the age, such as Henry Ford and the Wright Brothers, Carrier was continually looking for ways to create the next big must-have for society.

He found the inspiration right in front of his nose. As he stared through the mist, he realized that air could be dehumidified by passing it through water. Paradoxically, the moisture in the air would condense into mist droplets, thereby drying the air. If he could control humidity, what kind of impact could this have on human life?

The impact proved to be enormous. Humidity control became the basis of modern air conditioning. As a result, vast areas of the world are now bearable in the summer months. Imagine running a business in Phoenix or Dubai without air conditioning. Movie theatres, shopping malls, long-haul flights, and computer servers all benefit from the brainwave of Willis Carrier. And the company he established, bearing his name, prospers to this day.

Again, it is important to note that air conditioning isn't all good news. Air-conditioning units, along with electric fans, account for 10 percent of global electricity consumption and can leak harmful gases into the atmosphere. As regulations to control the use of refrigerants have become stricter, software has been developed to help businesses and public bodies track their refrigerant-related activities, ensuring compliance and lessening the impact of air conditioning on the environment.

By the way, the first home refrigerator was introduced in 1913, and then popularized by Frigidaire (owned by fast followers General Motors) in 1918.

Question 3

Which famous scientist said, "Chance only favors the prepared mind?"

A. Charles Darwin
B. Louis Pasteur

The answer is B: Louis Pasteur

Louis Pasteur was an incomparable Force for Good. The 19th-century scientist is perhaps best known for his germ theory, which demonstrated that microorganisms cause disease, something that was largely misunderstood until the development of high-powered microscopes. By *pasteurizing* (heating) wine (and then milk), he killed the bugs that quickly contaminate liquids and endanger individuals. Pasteur's discovery helped to save the French wine industry, which had been plagued by microbes that caused off-flavors and spoiling during fermentation. Pasteur then turned his efforts to immunology, making vaccines against fowl cholera, anthrax, and rabies by using the microbes in weakened states.

Charles Darwin, a contemporary of Pasteur and a great scientist in his own right, would likely have agreed with Pasteur's observation. His theory of evolution by natural selection was an inspiration borne of rigor, not good fortune. Darwin's studies of invertebrate animals as a young student at Cambridge prepared him to recognize the signs of evolutionary development in the plants and animals he observed during his five-year voyage on the HMS *Beagle.* Then, like many entrepreneurs, Darwin had to wait until the timing was right to go public with his ideas in his book *On the Origin of Species*. He sat on his bombshell for two decades until the culture of Victorian Britain had evolved to cope better with its revolutionary implications.

Question 4

Which company invented the digital camera?

A. Canon
B. Kodak

The answer is B: Kodak

A lthough most people don't realize it, digital photography was originated by a company generally associated with traditional film-based photography. Kodak engineer Steven Sasson developed a prototype digital camera in 1975. Weighing in at four kilos, the machine used a digital cassette tape to take black-and-white images at a resolution of 0.01 megapixels. Today's smartphones have a 12-megapixel camera and weigh less than 180 grams.

In its day, Kodak had ruled the world of photography. Founded in the late 1880s, the U.S. company was a notable pioneer in the 20th century. The "Kodak moment" passed into everyday language, while taglines like "You press the button, we do the rest" were wildly successful at making photography a pastime enjoyed by millions. Ultimately, Kodak also played a significant role in the evolution of the first commercial digital cameras, working with partners like Canon and Apple. Yet, despite taking a technological lead, Kodak ultimately failed to heed the changing culture in which it operated. Kodak missed the moment, allowing other companies to dominate the field of digital imaging.

Kodak filed for bankruptcy in 2012, but is now back in business, with a focus on innovative printing materials. Canon was a fast follower and now controls half of the world's digital camera market, although its sales have been impacted by the increasing sophistication of smartphone cameras.

Question 5

Which came earlier?

A. The first airplane flight across the English Channel
B. The first unassisted swim across the English Channel

The answer is B: the first unassisted swim across the English Channel

In 1875, Englishman Matthew Webb made history by becoming the first person to swim across the English Channel without artificial aids. Covered in porpoise fat for protection from the cold, he completed the 40-mile crossing just under 22 hours (suffering several jellyfish stings in the process). Earlier that year, American Paul Boyton had crossed the channel, but he had used an inflatable suit. Webb became a national hero, although his newfound celebrity status eventually went to his head. He tried to swim through the Whirlpool Rapids beneath Niagara Falls—despite warnings that any attempt was suicidal. His memorial reads, "Nothing great is easy."

Boats have carried people across the Channel for at least 3,500 years. A French air balloon floated over to England as early as 1785. In 1880, construction began on a tunnel that would eventually connect the United Kingdom to the continent in 1994.

Meanwhile, the first airplane to make the trip was piloted by French aviator Louis Blériot in 1909, complete with handlebar mustache. In just 36 minutes, Blériot was transformed from a bankrupt inventor into a world-famous aircraft manufacturer. His company Blériot Aéronautique was a leading French producer of planes until 1937.

Question 6

Why was Leonardo da Vinci's notebook hard to read?

A. He wrote backwards
B. He wrote in the Greek alphabet

The answer is A: He wrote backwards.

R enowned for paintings such as the *Mona Lisa* and *The Last Supper*, the Renaissance polymath Leonardo da Vinci was also a prodigious scientist, known for his studies in anatomy, astronomy, botany, hydraulics, and other fields. He was also an inventor who imagined extraordinary innovations—like airplanes, helicopters, parachutes, diving suits, and robots—centuries before technology caught up. Something of a perfectionist, his final words were, "I have offended God and mankind. My work did not reach the quality it should have."

Leonardo wrote in his notebooks using a shorthand of his own creation. He also mirrored his writing, moving the pen from the right side of the page to the left. Why? One theory is that he was left-handed and wanted to keep his pages—and his hands—clean of ink smudges. Others say his goal was to prevent rivals from reading his notes. But perhaps it was just more inventive that way!

1

INTRODUCING
SUSTAINABILITY AS
A FORCE FOR GOOD

- How Patagonia found its purpose
- Understanding the definition of *good*
- Business must fill the trust vacuum
- Why the Lions will rise . . .
- . . . And the Ostriches will die

"Every time I do the right thing, I make money."
—Yvon Chouinard, founder of Patagonia

T his quotation could have come from an Andrew Carnegie, a Henry Ford, or a Jeff Bezos. But it was said by the founder of the global outdoor clothing company Patagonia. And far from expressing self-satisfaction, it expressed Chouinard's conviction that the challenges of staying true to the company purpose have compelled it to change for the better.

Chouinard didn't set out to build one of the world's most successful brands. His first love was rock-climbing equipment, which he forged in his parents' backyard in California. He sold climbing gear to fellow mountaineers from the trunk of his car, raising enough money to subsidize his next trip to the Rockies.

Whenever the young Chouinard went into the mountains, he had ideas on how to make climbing equipment better than before. As a result, his fledgling business—not yet bearing the Patagonia name—was soon the leading supplier of climbing hardware in the U.S. But while his reputation as an engineer grew, the company's profit margins stayed tight. For one summer, Chouinard survived in the mountains on a diet of damaged-can cat tuna, "supplemented by oatmeal, potatoes and poached ground squirrel and porcupines."[1] His enterprise was starting to look far from sustainable.

Start with Values

In 1972, Chouinard launched Patagonia, with the goal of selling rugged clothing to the climbing and surfing community as a cash cow to feed his core manufacturing business. Today, the company nets profits of more than $100 million a year. Chouinard's determination to make things better than before drove Patagonia's expansion from the back of his car into a global champion of social and environmental stewardship with over 100 stores worldwide.

But while Patagonia is held up as a responsible, "do-good" brand, its ethos originated from a rebellious mindset. Chouinard looked at the typical corporate rules and standards and decided he wanted to break them. Why run his business the same as everyone else?

Chouinard's leadership team, none of whom had a business degree, shared his attitude. So they started from their personal values. For mountain gear manufacturers, quality was the watchword. When you're relying on your equipment to keep you safely anchored to a sheer vertical rock face, the smallest defect could lead to accidents or death. But what did quality mean in apparel retail? There were no books to read from, so they wrote their own "philosophy of quality." Little by little, they began to define what quality would mean for Patagonia:

> Number one . . . we had to have multi-functional clothes, because we didn't want to own a lot of clothes. We wanted to have a ski jacket that . . . you can wear on top of your suit coat in a rainstorm in Paris in middle of the winter. Another criteria for quality was don't chase fashion. We didn't want to make disposable clothes, and we wanted to cause the least amount of harm in making those clothes.[2]

Years later, Chouinard summed it up in an interview this way: "The best thing you can do is buy the best product you can and keep it going as long as possible."[3]

And forget the corporate uniform or attendance rules laid down by a human resources department. Patagonia is known for a culture in which employees have been known to go surfing or powder skiing during work hours, and to come to the office barefoot or with their kids in tow—as long as they do a good job. Over the last half century, Chouinard himself has been famously hands-off as a leader, encouraging deep-seated independence among colleagues that results in boundless loyalty and also a company-wide desire to innovate.

"All I care is that the work is done," he said. "And so to do that, you have to hire self-motivated, very intelligent people who know their job, and then you leave them alone."[4] The result is a business model that attracts good people. For every job opening, Patagonia has an average of 900 applicants, allowing the company to pick and choose from the very best.

Refuse to Do the Wrong Thing

Vincent Stanley, an original employee at Patagonia, cowrote *The Future of the Responsible Company: What We've Learned from Patagonia's First 50 Years* with Chouinard. In the book, he explains the company's evolution into a purpose-led business:

> We did not set out to be a responsible company, but time after time we stumbled into virtue after discovering we were causing harm. As climbers and surfers, our direct engagement with nature allowed us to recognize the environmental crisis earlier than others. For us it was clear that the health of nature undergirds the health of our social and industrial systems. . . . Every time we refuse to do the wrong thing, the constraints we place on ourselves force innovations that result in products we otherwise would not have developed.

And with each new innovation comes new customers and a stronger reputation.

Today, sustainable companies rely on techniques like life-cycle analysis, ecosystemic thinking, stakeholder management, and supply chain risk management to understand and manage their environmental and operational impact. Although it may have used different terms, Patagonia has adopted such methods for decades. The company is a leader in understanding the impacts of its operations on people and the planet.

As the world's environmental problems grew more and more complex, Patagonia's leaders recognized that clothing retailers were part of the problem. In 1991, the company commissioned a study to assess the environmental impacts of its four most commonly used fibers: cotton, polyester, nylon, and wool. Cotton was the villain in the pack. The company was so appalled at the environmental impact of intensive cotton farming that it decided to take all its cotton sportswear 100 percent organic by 1996. Patagonia now aims to exclusively source regenerative organic cotton and hemp that returns precious topsoil to farmland and traps carbon dioxide.

One day, it's possible that the company will use only recycled and renewable raw materials. For example, suppliers have developed 100 percent recycled polyester with the same performance and durability as new polyester made from freshly drilled oil.

Patagonia launched a website known as the Footprint Chronicles in 2007 to "bring transparency to its supply chain and to tell the stories other companies typically don't tell."[5] Visitors to the site can easily learn about their suppliers' working practices, including an interactive map with their locations. The company did not own the farms, mills, or factories, but the work was done in its name, so Patagonia took responsibility for their practices. Investigations into the supply chain shined a light on industrial processes, safety measures, air quality, and factory wages. To reduce environmental and operational risks, Patagonia cut its factory count by a third and reinforced its relationships with the remaining partners. Today, a member of the social environmental responsibility team will visit a new factory to verify conditions before placing an initial order—backed by the authority to veto the deal if they are not fully satisfied.[6]

In partnership with Fair Trade USA, Patagonia today makes clothes with an assurance that some of the money spent on a product will go directly to local workers and stay in their community. Democratically elected committees decide how the money is spent.

"Don't Buy This Jacket"

Patagonia is also a leader in the effort to reduce needless waste. The company will accept any worn-out Patagonia product for recycling or repurposing. Worn Wear, Patagonia's used clothing and repair program, has grown from a mobile repair truck into the largest garment repair facility in North America. In 2011, the company ran an inspired ad in the Black Friday edition of the *New York Times* bearing the headline "Don't buy this jacket," beseeching customers to avoid purchasing clothes that they didn't need. On the face of it, selling fewer products is bad for business. That is, unless your quality products attract new customers who share your belief in the preciousness of nature.

The list of accolades and industry-firsts that Patagonia can boast is extensive. Patagonia partnered with Walmart to create the Sustainable Apparel Coalition, which now includes some of the world's biggest clothing retailers and provides a consumer-facing index that shares the social and environmental impact of a garment.

The company has taken a lead on recycling materials, such as polyester fleece made from soda bottles or hat peaks from abandoned fishing lines. It has eliminated dye colors that required the use of toxic metals and sulfides. Patagonia also partnered with Samsung to develop a washing machine that captures microfibers shed from fleece, which is made from oil-based polyester.

Patagonia's widespread use of recycled materials contributed to LEED certification for the expansion of its Reno, Nevada, distribution center, reflecting the facility's eco-friendly use of energy and other resources. When it needed a new distribution center in the east of the U.S., Patagonia eschewed undeveloped land to site a 360,000 square-foot warehouse on the site of an abandoned mine.

In 2012, Patagonia became the first company in California to be certified as a B Corp, meeting high standards for social and environmental performance, transparency, and accountability.[7] In 2019, Patagonia was named a U.N. Champion of the Earth for putting sustainability at the heart of its successful business model.[8]

Patagonia also donates one percent of sales—profitable or not—to environmental causes. By founding the 1% for the Planet alliance, it has brought together over 4,870 business members in 110 countries that also pledge one percent or more of their annual sales to over 725,000 environmental partners. Then there is Tin Shed Ventures, named after Chouinard's original blacksmith shop, which funds startups that place environmental and social returns on an equal footing with financial returns.

"Everyone in Patagonia knows to take one step toward responsibility, learn something, then take another step," says Vincent Stanley. "Many of our suppliers and customers have also become invested in this process of improvement."[9]

Driven by Purpose

In 2018, Patagonia changed its purpose statement to "We're in business to save our home planet." Chouinard wanted to demonstrate a mindset shift that acknowledged that reducing the impact on the planet is not enough. The urgency of climate change meant that now was the time to start healing it.

In 2022, the Chouinard family took the extraordinary step of making the Earth the company's sole shareholder. Every dollar that is not reinvested is now distributed as dividends to protect the planet. "Instead of extracting value from nature and transforming it into wealth, we are using the wealth Patagonia creates to protect the source," said Chouinard. "I am dead serious about saving this planet."[10]

Patagonia has become the epitome of a responsible company. For Stanley, it's a quality that every business will need to emulate. "What's certain is that any 21st century business seeking to keep customers and make new friends will need to improve the environmental and social performance of its products," he wrote. "More customers will demand to know: Does your product or service hurt them or their children? Does your product hurt the workers who make it, or their community, or the ecology of the place where your components are drilled, mined, farmed or stored? Is your product worth its social and environmental cost?"[11]

Unleashing a Force for Good

Patagonia's journey is evidently successful on a commercial level. But I would argue that the company has become so successful by proving itself a Force for Good. Had the ambition stalled at producing a reliable jacket, then Patagonia might well be yesterday's news. Instead, its purpose provides inspiration to millions of customers around the world.

Patagonia shows that sustainability as a Force for Good presents a huge opportunity for long-term growth. It's an approach that requires both innovation and tenacity. As Jim Collins wrote in *Good to Great*, "Greatness is not a function of circumstance. Greatness, it turns out, is largely a matter of conscious choice, and discipline."[12]

The demands placed on businesses by customers, communities, stakeholders, and employees today have never been higher. CEOs are facing increasing pressure to demonstrate that their enterprises offer solutions to the world's unprecedented problems, including social pressures such as unsafe work conditions, in addition to environmental issues like hazardous emissions, water shortages, soil erosion, mineral extraction, plastic mishandling, deforestation, pollution, and waste. Every company needs a license to operate from society at large, and that's no longer guaranteed by a healthy balance sheet.

Those businesses that fail to navigate a growing loss of trust among consumers and talented employees will join the list of casualties. Those that are blindsided by government legislation, or suffer deep-set reputational damage, will also decline. Doing the right thing requires more than adding a colorful section to the annual report. It's a commercial necessity. Real values drive real value.

The constellation of stakeholders, as mapped by Klaus Schwab, founder of the World Economic Forum, offers an easy-to-grasp image of the complex social and environmental challenge today's businesses face because of the diversifying array of stakeholders they are required to satisfy (next page).[13]

This existential challenge also presents an exciting opportunity. Those businesses that can retain the trust of society and find the latitude to think long-term stand to grow sustainably in the years and decades to come. More than simply surviving, they will truly thrive.

The need for resilience has grown during the rapid changes that have taken place in the 21st century, with more uncertainty to come. In 2019, who could have predicted the seismic impact of Covid-19 around the world? In its

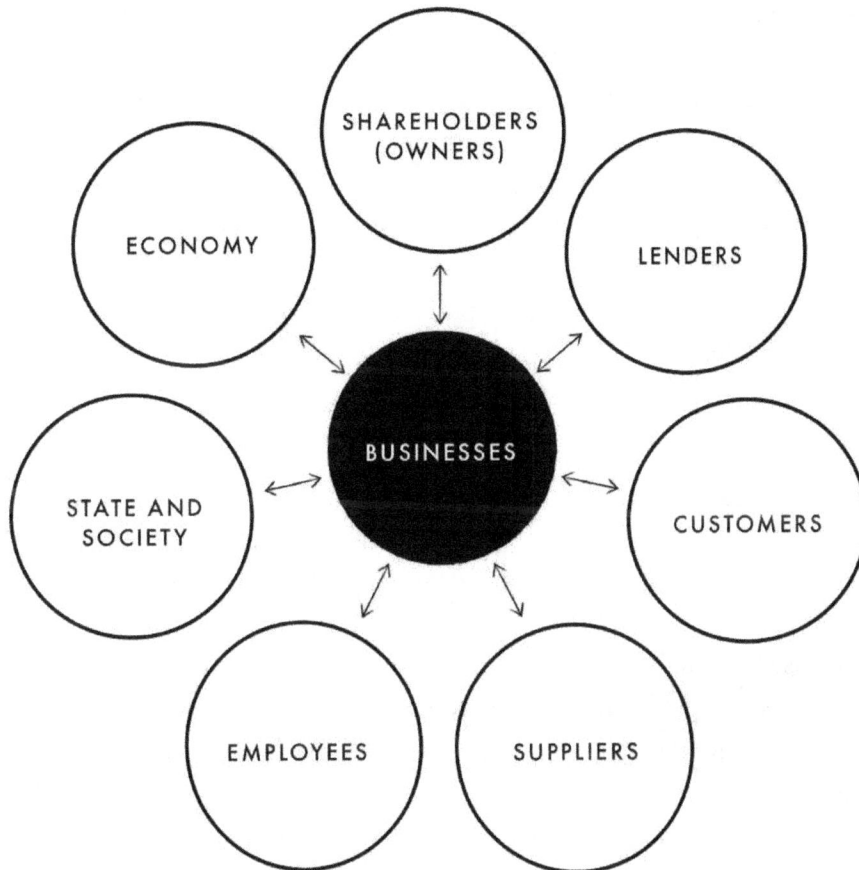

THE CONSTELLATION OF STAKEHOLDERS

most recent quadrennial analysis of global trends, the National Intelligence Council of the U.S. envisions a world that will face more intense and cascading challenges that will "repeatedly test the resilience and adaptability of tomorrow's companies."[14]

But while innovation and growth are essential for survival, they are hard to achieve. Just one out of every nine companies maintains profitable growth for a decade or more.[15] And corporate performance is more volatile than ever.

On average, a third of publicly listed companies cease to exist every five years—an attrition rate that's five times higher than it was 50 years ago.[16]

What's the antidote to growing instability? Businesses must prioritize organizational agility to respond faster than their competitors—and that includes creating a clear line of sight with all stakeholders, whether customers, employees, suppliers, communities, or the environment. CEOs still aim to maximize revenue and returns, but the notion of value is now different. What's good for the stakeholder is, by virtue of shared interests, also good for the shareholder.

Businesses as Trusted Institutions

Fortunately, businesses have already done much to earn the trust of the societies in which they operate. Free-market economies over the past century have raised the quality of living, per capita income, and life expectancy for billions across the globe. In the 1950s, Norway stood out as the most long-lived nation, with an average life expectancy of 72 years. That's now the global average.

The Edelman Trust Barometer, which measures trust levels in institutions around the world, reveals an "epidemic of misinformation" that has created an "environment of information bankruptcy and a mandate to rebuild trust and chart a new path forward."[17]

The decline in societal confidence across all institutions means that business is now the most trusted institution, according to the Trust Barometer, with even greater expectations placed upon it.

Importantly, two-thirds of investors are more attracted to businesses that align with their own values. There's an inherent logic to this reality. Publicly-owned businesses are answerable to the pension funds that represent firemen, teachers, and nurses—which will only invest in brands that reflect the values of firemen, teachers, and nurses.

Employees are another stakeholder group for whom personal values are deeply motivating. Surveys show that over 90 percent of people under age 30 agree that the more their companies become mission-oriented, the more their motivation and loyalty will increase as employees.[18] Three-quarters of LinkedIn members say they want jobs that offer a sense of purpose.[19] In companies that clearly articulate how they create value, twice as many employees say they're motivated to perform compared to the employees of more opaque organizations.[20]

Younger people, whether as customers or employees, are especially concerned by the environmental and social footprints of brands and businesses. Eighty-five percent of Gen Z customers tell others if they have had a positive experience with a "good" company, while 83 percent of Gen Z in the U.S. say that a company's purpose is an important factor when choosing an employer.[21]

Having a Noble Purpose to guide the company forward in a positive direction does more than attract new talent. It can become a powerful strategic tool for sustainable growth, providing a mandate to improve over time. In periods of uncertainty, a Noble Purpose will also give companies a starting point and ready guardrails for making rapid decisions.

Yet, while almost one in nine global executives understand that purpose is important, fewer than half integrate purpose into strategy.[22] Worse still, fewer than half of employees know what their organization actually stands for and what makes it different.[23]

However, those businesses that get the purpose challenge right find that they deliver revenue growth, ongoing transformation, and innovation as a result. In too many cases, there is a looming "purpose gap" between the good intentions of leaders and the everyday actions taken throughout a business— but on the plus side, the gap can be filled by improved behaviors that will energize long-term growth.

Companies have an opportunity to make money and do the right thing. Firms that show both high purpose and management clarity achieve higher financial and stock market performance in the future.[24] And when employees believe in their organization's purpose and understand how to bring it to life, so their own performance will improve.[25]

This long-term approach is not always easy, but it has been shown to work. In a study that looked at 615 large and mid-cap U.S. publicly listed companies from 2001 to 2015, the McKinsey Global Institute concluded that those with a long-term view outperformed the rest in earnings, revenue, investment, and job growth.[26] They have proven more resilient to economic crises.

Firms that perform well on material sustainability issues "significantly outperform their counterparts over the long-term on both the stock market and financial results," according to professors at the University of Oxford, London Business School and Harvard.

The Lions Roar—Then Eat the Ostriches

In the early 1990s, I joined a startup company that recycled the electronics from computers and telecommunications that would otherwise be headed for landfill. Our business premise was simple: Why damage the environment with this toxic waste when it could be sold to people who need it? The reusable parts were refashioned into computers for communities in Africa and Asia.

In making this move, I was turning my back on a safe career path in a reputable law firm. One day I was working in a big, shiny office in downtown Chicago, all suits and ties. The next, I was in a T-shirt and hard hat, surrounded by smokestacks. In my mid-20s, with a young family, it was a risk. But the work felt real. I could see the impact through my safety glasses.

Besides, this was personal for me. My parents had fled persecution in occupied Ukraine and emigrated to the United States, where they found jobs in the factories of Chicago. They made their home in the city that's proud to be

> Hog Butcher for the World
> Tool Maker, Stacker of Wheat,
> Player with Railroads and the Nation's Freight Handler;
> Stormy, husky, brawling,
> City of the Big Shoulders

(from "Chicago" by Carl Sandburg)

My parents, like so many other immigrants, were thankful for a new start. They valued the opportunity to contribute in a free society through meaningful labor. Yet it's also true that the factory shifts were unrelenting. They experienced working conditions that were both unsafe and environmentally harsh.

Theirs was the Chicago of Upton Sinclair's novel *The Jungle*, where from "the line of buildings . . . rose the great chimneys, with the river of smoke streaming away to the end of the world."

Growing up, I had seen how this work had affected them. While the U.S. has come a long way since the 19th century, I believed that more could be done to create a safer, more sustainable, and productive world—and that business could lead the way. I was determined to make a difference, starting with my small recycling enterprise.

Back then, it was already clear that some companies were ahead of the environmental curve. They could sense the direction of travel. Having their picture on the front page of *The Wall Street Journal* for dumping computers would be bad for business. They'd come to us and ask, "How do we recycle the glass from our monitors? What do we do with the tons of rubber from our cabling?" So we innovated and provided them with solutions.

This was stakeholder management before the term existed. In the early 2000s, I then developed an algorithm implemented through a software program that demonstrated the concepts of "value to the customer" and "value from the customer."

On one side of the spectrum, some corporations want to maximize value from their customers. They seek to extract as much as possible from stakeholders, whether by paying low wages or by sourcing from suppliers that pollute the environment and damage communities. They may make money in the short term. But the greater the profit margin, the worse the product or service, and the worse their reputation. Their business is unsustainable and will fail. The board might even be sent to jail.

On the other side of the spectrum, there is the outlier customer who wants a cheap product that has all the bells and whistles yet earns minimal profit for the corporation. Unless the customer pays a fair price, the corporation must drive efficiencies at the expense of its stakeholders, such as through low wages and poor working conditions for employees. Again, this is unsustainable, and the business will eventually fail.

Using a stochastic dashboard designed to analyze widely varying statistical patterns, I demonstrated to Fortune 500 companies all over the world how to balance value in and value out in a way that yielded an upward growth trajectory over the long term. Importantly, the tradeoff model included all stakeholders. I invited CEOs to dial up and down variables such as workers' wages or the reputational damage of an environmental disaster. I'd show the likely effects of new regulations or a customer scam.

The takeaway was straightforward. If you're interested in lifetime customer value, rather than a one-off transaction, then making decisions with potentially negative impacts isn't worth the risk.

I discovered that these were the champion corporations –let's call them the Lions—that immediately saw the need to go above and beyond. Their CEOs recognized the dangers of inactivity in the face of social and environmental challenges. They wanted to be at the forefront, leading the change.

Then there were those who attempted the least-amount-necessary approach: the Ostriches, with their heads stuck in the sand. Their mantra: "What's the bare minimum we need to do for compliance? How do we stay on the right side of the law? That's all our shareholders expect. We'll make no changes until we have no other choice."

When the Great Recession hit in 2008, most firms went into survival mode. I saw that the Lions came out of the slowdown stronger. Those forward-thinking corporations were ahead of the game when regulations subsequently came into force. The Ostriches had to play catchup, and many are no longer operating today. Some were even eaten by the Lions.

What Is Good, Anyway?

The definition of *good* has taxed great thinkers for centuries. For example, the Greek philosopher Aristotle believed that practical wisdom was the highest intellectual virtue and therefore the greatest source of good.

Candace Vogler, professor of philosophy at the University of Chicago, translates the notion of practical wisdom for modern businesses. "It's really about finding balance, and that's useful no matter what you do," she says. "Wisdom may sound very elevated, but it's basically a matter of thoughtful decision-making and choice implementation."

Vogler adds, "In business, the problems that people run into are often caused by short-sightedness under pressure. They might ignore a lot of stakeholders because it's inconvenient to take their interests into account."

She believes that an ethically sustainable business will develop close relations between management and employees, between the firm and the community it's involved in and even with its competitors.

Exit the Ego System

Vogler recommends that firms and their people consider what kind of good they are actually providing to the world. "What does it mean to participate in producing and sustaining that good? For whom are you providing this good, and what are the perils involved? What are the lines you won't cross?—And what will you do if other people are behaving unethically?"

The first step to practical wisdom, in Vogler's view, is to leave your ego at the door. "It takes courage to be selfless. But if you can't figure out how to get over yourself, it's going to come back on you in a bad way. You want to be the right size as a human among your fellows."

Vogler also teaches students about self-transcendence: the goodness that comes to people from belonging to something bigger than just themselves, such as a faith, a family, or a social purpose. "Again, you can see that sentiment hold true at sustainable businesses, where everybody—from the

CEO down—is in the service of a larger good that reaches beyond individual advantage. When lots of people buy into that good and share commitment to the firm's longevity, then it makes sense that the firm will thrive."

The Only Way Is Sustainability

I'm now CEO and founder of Sphera, a global operational and sustainability risk management software and data services company that exists to catalyze sustainability and safety programs by providing a foundation of robust, specific, and measurable data and expertise on which companies can build positive change. Our software, data, and consulting services help a company, wherever it may be on its journey, to become a greater Force for Good by operationalizing sustainability.

Let's be in no doubt: The transition to a more sustainable future is already happening. Regulatory intervention is now widespread, and the pace is only going to increase. The cumulative number of sustainability-related policy interventions per year has accelerated from zero to over 500 in the last 40 years.[27] They have more than doubled in the last decade, while money allocated to sustainability-related investments has increased 25-fold since 1995.[28] In total, more than $30 trillion is invested in sustainable assets globally, according to calculations by the Global Sustainable Investment Alliance.[29]

To take just one example: For too long, businesses shelved the need for action on sustainability, and the results of that inaction are clear to see today. Around the start of the 21st century, consumer sentiment began to turn, prompting many organizations to draft strategies that would create enough noise to keep society and governments on their side, but demand the least amount of effort. It's now clear that such minimal efforts are not enough.

Over the last decade, sustainability has become a major concern of risk management for corporate boards, who recognize the potential for reputational damage and subsequent loss of customers, investors, and employees.

Falling short can now bring rapid consequences. When Deutsche Bank was raided by prosecutors in 2022 over charges of greenwashing—phony claims to environmental sustainability—the CEO lost his job and the share price tumbled by 24 percent.

Sound intentions aren't enough. Ticks in boxes are simply that. Stakeholders increasingly expect sustainability models that are operationalized. Our efforts need to be measurable, transparent, traceable, and auditable—with a 360-degree view of objective sustainability metrics.

As sustainability has risen up the agenda, so companies are recognizing the need to make real change and turn lip service into hard facts. They are under pressure from governments, investors, and customers to show how sustainable they really are. Increasingly, they are expected to shine a light on their supply chains, too.

Today, about half of the world's companies have made sustainability commitments, but only a fifth say they have a clear roadmap for incorporating sustainability strategy into their core business, according to a survey by Sphera.[30] Firms know they need to enact change, but finding the right solution that will practically aid them on their sustainability journey is proving a challenge.

Every organization must accept that this is no passing craze. As sustainability becomes standard operating procedure, so it will impact every business decision exponentially. In the near future, an enterprise sustainability report will be just as important to investors as financial metrics.

It's therefore vital that organizations begin planning for the long term. That foresight will decide not only which organizations will survive the 2020s—but also which of them will thrive.

Sustainability Stems from Abundance, Not Scarcity

Deborah Cloutier is chief sustainability officer at Legence, a global provider of energy efficiency and sustainability solutions for the built environment. Backed by three decades' experience in this specialist field, her advice helps clients to reduce energy costs and meet decarbonization goals while improving the health and wellness of their buildings.

One of Cloutier's several claims to fame is that she worked on the first-ever greening of the White House, improving the energy and environmental performance of the whole complex as part of a team of architects and engineers who went on to found the U.S. Green Building Council.

Cloutier insists that sustainability is about more than defense. Every day, she sees the financial and competitive advantages that stem from a proactive sustainability mindset. She says:

> It's critical that you start from a robust business case, rather than trying to dress up things you're already doing. Think of sustainability as coming from a place of abundance, rather than deprivation or scarcity. It's an opportunity set and a source of ROI. Why? Because energy efficiency pays. It always has and it always will by reducing your operating costs.
>
> Then there are the secondary financial benefits of green amenities and improved air quality, such as incremental asset value; better health, well-being, and safety; increased productivity; and talent attraction and retention. These aren't "nice to haves"—they are pure business plays.

Sustainability partners like Legence are riding some powerful tailwinds, as businesses recognize the need to control costs and make facilities more efficient.

Cloutier highlights the global surge in demand for digitization as another major opportunity. The world can't build data centers fast enough to meet the need for cloud computing, Generative AI, and storage solutions, which are all dependent on high energy provision. Data centers could require nine percent of the annual energy generated in the U.S. by 2030, more than doubling their current consumption.[31]

"These data centers need to be scalable and high density, calling for innovative design, prefabrication, and retrofitting, which plays to our strengths at Legence," says Cloutier. "Healthcare is another sector that is invaluable and energy intensive. There's also a megatrend to reshore manufacturing within the U.S., borne out of Covid-19 and supply chain issues. Add to that the growing demand for so-called 'green collar jobs,' particularly in the renewables space."

More widely, capital providers—especially European financial firms—are mandating a sustainable approach. So organizations have no choice but to change the way they're doing business. Any business that wants to raise capital will need a good GRESB score, which measures the sustainability of a company's real estate practices, as well as a sound decarbonization strategy.

"Business leaders are also becoming more climate conscientious," adds Cloutier:

> As individuals, they want to sign onto decarbonization frameworks or net-zero commitments, so they need advice on strategy, design, and construction. CEOs are looking out their window or watching the frequency and intensity of weather events and wildfires, and they're drawing their own conclusions on climate change. They want to make their buildings more resilient, whether it's a new construction or a retrofit. Regulation around the world is only increasing that momentum.

Back to the Future of Risk Management

Cloutier describes the current sustainability sector as going *back to the future*:

> When I started out, the emphasis was on energy efficiency, utility cost reduction, and optimization. We didn't talk a lot about carbon. Sustainability then evolved into a vessel for communicating values and laying out a softer set of responsibilities.
>
> Today, we're back to sustainability as a risk management strategy that is much more embedded into an organization's decision

making. In fact, our clients talk about *climate risk management.* It's not CSR or ESG. Sustainability has grown up and grown teeth. Reporting regulations bring legal obligations and monetary fines for noncompliance. It's now a standard rather than an aspiration. It has metrics and proof points, like the other elements of risk management. Sustainability is about cost, not hugging trees.

Over the last 30 years, Cloutier has seen incredible headway with the normalization of the triple bottom line and the sheer range of building solutions that are available today. But she still gets frustrated at the industry inertia and lack of accessible funding that has slowed the rate of transformation:

> No one company or organization is going to fix those barriers overnight. It's going to take whole armies of engineers—building, electrical, plumbing, mechanical, civil, you name it—to figure this out. The truth is that humankind has never attempted to do what we're trying to do right now, which is take existing building stock and decarbonize it deeply. But then you see how far we've come in the U.S., and you think how much efficiency we've already banked; you realize that anything is possible.

Business Will Find a Way . . . It Always Does

Individually, we have never had so much power to influence the companies that provide our everyday products and services. Smart companies see societal change before it comes. Those laggards that wait for regulation will already be winnowed out by the time it arrives.

CEOs still want to maximize shareholder profit—that's all well and good. But they can't do that today without realizing there are other inputs into the system, such as employees, investors, the environment, and communities. When they get that right, the prize is long-term shareholder value.

I just can't see how the Ostriches will survive. Market forces will take over, and they will die out. People will not want to work for them. People will not want to buy from them. People will protest and tell their friends and social network connections to shop elsewhere. Companies aren't immune to the will of the people.

Those businesses that get ahead of stakeholders—or at least listen to them and respond to them—will increase shareholder value in the long run. Businesses that fail to navigate the growing loss of trust among consumers and talented employees will add their names to the roll of mergers, acquisitions, and bankruptcies. Those who think they can outsmart their stakeholders and maximize short-term profits won't be around very long.

Yet I'm optimistic for the future. When you look at all the struggles that humanity has faced, we've always found a path to the other side. Whatever challenges we face, business will find a way, as it has always done before. Society pushes businesses, and businesses push themselves, and eventually, the winners make it happen. Innovation is a remedy. Pioneers will emerge. It's not without its ups and downs. It's never perfect. There's so much we still need to do. But in the long run, it will get done.

How to Make Sustainability Work for Your Business

Since 2020, I've seen businesses take to sustainability in a manner I could never have predicted. I truly believe that operationalizing sustainability is the only way to create a cleaner, safer, and better world. It will become the new normal in company performance.

Right now, many companies' sustainability efforts are often too broadly described, vaguely promised, and lacking in action. They must be refined, defined, and monitored to be effective. For companies looking to drive genuine positive change and build a better world—as many are—there is no standardized way to report and measure progress.

This book will help you find the missing pieces of the sustainability puzzle: the expert insights, data, and technology that are needed to catalyze authentic, measurable, positive change. It offers a framework to gauge your current status, think big (but realistic), and take manageable steps forward on the journey.

We'll explore the dynamics at work behind being a Force for Good. What does it mean for today's businesses? And *how* can your business become a Force for Good in society, for the long-term benefit of all your stakeholders? Meaningful progress is within reach.

Key Takeaways

- Pioneers like Patagonia demonstrate that acting as a Force for Good is beneficial for both business and society.
- Now is the time for business to lead.
- Be warned! The influence of individuals in society—customers, employees, and investors—will determine the success and failure of companies.
- CEOs broadly welcome the need for scrutiny on sustainability, but they often lack the necessary information and frameworks to respond. This book aims to fill that gap.

2

THE CYCLE OF GOOD

- Cultural change is a catalyst for transformation
- History shows that good and creation are connected
- Businesses that sense this cycle are rewarded
- Those who ignore it will decline
- Good companies balance value to and from their stakeholders

"And God saw all that He had made, and behold, it was very good."
—Genesis 1:31

Good inspires innovation. History shows there is an essential link between the two at a societal level, both for early civilizations and in the modern day.

Take the story of creation from the Bible. Over six days, so we're told, God created light, the earth, land and vegetation, the sun, moon, and stars, all the fish, animals, and birds, and then finally man and woman. On the seventh day, God rested. He looked back over his handiwork and regarded it as "very good."

The creation story in Genesis was likely written around the sixth century BCE, although the words would have been passed down orally for many generations before that.[1] Those Hebrew scribes and their audiences recognized an integral relationship between creation that enhances life (light, water, earth, vegetation) and goodness. "Be fruitful, and multiply, and replenish the earth," the Bible continues (Genesis 1:28), signifying the special opportunity that God gives people to use their creativity for the greater good. The very nature, and one of the main purposes, of humans is to be creative.

Humanity is also given "every herb-bearing seed" and "dominion . . . over every living thing that moves on the earth." The writers of scripture lived in a world where communities saw themselves as stewards of the land, responsible for making it a better place to live for all. They celebrated a balance between nature and humans, sparked by innovation, which goes beyond "good" to reach "very good."

Every known language has a word to express the idea of good (and bad). The distinction between right and wrong seems ingrained in human societies, even though the line between the two sometimes shifts to reflect contemporary needs and philosophies. What constituted good in the past was often far removed from modern values.

The link between creativity and good in society resurfaces throughout recorded history up to the present day. It has played a central role in maintaining social order and harmony. Good has proved both a civic necessity and a vital force for the balance of nature. In its name, extraordinary feats of innovation have pushed civilization forward.

Breaking the Seal of Knowledge

The invention of the printing press around 1440 in Germany by Johannes Gutenberg was arguably one of the most important innovations of the past 2,000 years. It brought the opportunity of education to the masses. Driven in part by the new, broader access to knowledge of all sorts, scientists, artists, writers, and rulers began to revisit the definition of *good*. Thinkers looked beyond religious texts to evidence-based reason and logic for their answers to questions about nature and humanity.

By the middle of the 18th century, Britain offered the necessary cultural context for the First Industrial Revolution. The culture was ripe for invention. Compared to its continental neighbors, Britain had enjoyed a period of internal stability and continuity. The scientific revolution had turned chemists and physicists into the rock stars and influencers of their day, hastening the development of the steam engine and other powerful transformative inventions. Facts were now prized above tradition and superstition. Patents were granted by Parliament to encourage and protect innovation. The London Stock Exchange opened in 1773, offering a flood of capital for investment in new ventures.

Coal was plentiful, and an innovative canal system—soon followed by the first railroads—allowed rapid nationwide transportation of raw materials that were flooding in from the empire. The world was getting smaller and moving faster. Changes in agriculture had shaken up the countryside, producing more food, and also an exodus of families who sought work and shelter in the cities.

For many people in Britain, increased production and efficiency from industrialization meant lower prices, greater choice of goods, and improved wages. Their standard of living greatly improved.

Of course, the Industrial Revolution wasn't all "good." As workers rushed to the mills, factories, and mines, the working conditions for women and children became a blight on society over the next century and beyond. Overcrowding, rising crime, poor urban sanitation, pollution, poverty, and prisons were all challenges that Victorian society took too long to address.

America Rising

Across the Atlantic, the U.S. would become the natural setting for the Second Industrial Revolution in the second half of the 19th century, emblemized by the expansion of the railroads. This era saw the rise of the telegraph, electrification, and petroleum as a new source of energy, in addition to the emergence of new forms of organizational governance needed to operate large-scale businesses across large areas and fast-growing cities.

The cost-effective production of steel is often held up as a trigger for the Industrial Revolution in the U.S., yet equally powerful was the spirit of peace, hope, and ambition at that time. Millions of immigrants were arriving from Europe in search of a new life. There was a surplus of capital for investment. This was a period of *laissez-faire* politics when individuals were encouraged by the state to pursue their own goals for the benefit of society. These economic conditions helped encourage Thomas Edison to invent the incandescent lightbulb in 1879, extending the working day and forever separating light from fire.

Innovations in transportation and communication linked isolated communities together in a process that cultural historian Stephen Kern has termed "the annihilation of distance."[2] Political, economic, military, and private life would never be the same. The telephone, invented by Scottish-born engineer Alexander Graham Bell, arrived as a miracle device in 1876, enabling people to "raise money, sell wheat, make speeches, signal storms, prevent log jams, report fires, buy groceries, or just communicate across ever-increasing distances."[3] By 1914, there were more than 10 million telephones in operation across the U.S.[4] Bell went on to cofound the American Telephone and Telegraph Company (AT&T), which remains one of the world's largest telecommunications companies.

By the start of the First World War, the United States had greater industrial output than Great Britain, France, and Germany combined. The Second Industrial Revolution had brought its own challenges. Journalists and authors decried the poverty and dangerous working conditions suffered by many. Yet at the same time, new technologies were raising the living standards of working-class families and growing the middle class to new levels, which in turn created a greater demand for goods. The stock market allowed a wider distribution of business ownership.[5]

The Consumer Revolution

The next golden age of innovation, arriving in the years of rapid growth following the Second World War, also mirrored the culture of the day. The hardships of the Great Depression and the world war gave way to a period of prosperity. During the Second World War, consumers had saved money in anticipation of a post-war bonanza. By 1945, Americans were stockpiling over a fifth of their disposable income on average, compared to just three percent in the 1920s.[6] When rationing stopped in 1945, the response was inevitable. Unemployment had fallen from as high as 25 percent to less than two percent in the previous 15 years.[7] Now families could afford a home and the labor-saving appliances to fill it. They could own a car and go on vacation.

This was the time of the baby boom and the suburban nuclear family. The major brands that emerged in midcentury America typify a culture of convenience and wholesome values. The American Broadcasting Company (ABC) expanded from radio into television broadcasting as ownership of TV sets rocketed from a few thousand to over 12 million by the early 1950s. (In the U.K., television ownership spiked in 1953 when Queen Elizabeth II's coronation was broadcast by the BBC.) The Best Western chain of hotels put down roots as families bought cars and went traveling. Fast-food giant McDonald's launched its Speedee Service System in 1948, while Carl's Jr. opened its first full-service restaurant in California. Dick's Sporting Goods began selling supplies to weekend fishermen in New York State. Fender was founded in 1946, designing guitars that would change the face of music in the following decades. IAMS offered some of the first commercial dog and cat food, while Minute Maid took its experience of providing orange and lemon juice to the military into the consumer market. Yet another iconic brand of that era was Tupperware, launched in 1946 by Earl Tupper. He saw an opening for airtight containers using improved techniques of plastic production. Advances in refrigeration helped Tupper's cause, as did innovative direct marketing in the form of popular Tupperware parties.

This was also a golden age of advertising, fueled by television, self-service retail, and supermarkets. To appeal to the vast and still-expanding mass market, brands needed to stand out from the crowd and tell stories. Beer companies, still traumatized by the Prohibition era, reinvented themselves as mainstays of a wholesome family life.

Christian Dior's New Look clothing hit the streets and magazines in 1947, with its nipped-in waist and full-skirted silhouette capturing a sense of femininity and luxury that was instantly popular after the years of austerity, minimalism, and a "make-do-and-mend" mindset. Estée Lauder also launched soon after the war, growing rapidly with a range of cosmetics that promoted individuality. For nearly 80 years, the brand's unstinting focus on the changing world of its customers has allowed it to stay ahead of their needs and widen its portfolio of products.

New car sales multiplied fourfold in the decade following the war, and by the end of the 1950s, three out of four American households owned at least one car.[8] Federal Housing Administration (FHA) loans and the GI Bill gave many veterans access to low-interest mortgages and the means to buy a home.[9] Firms like Levitt & Son in New York applied mass-production techniques from car manufacturing to build houses and entire communities at a rapid pace, in response to the twin forces of overcrowded cities and a migrating rural population. Washing machines and dryers, dishwashers, and garbage disposals freed up hours in the day for reading, education, and leisure.

Major industrial companies piggybacked on the American Dream, running extensive advertising campaigns that celebrated the development of new materials and new products. In some ways, these postwar years were a carefree time. Fears of environmental harm from substances like plastics were still in the future, even as pollution was commonplace and often devastating to biodiversity.

Thus, with the end of the Second World War, the U.S. repurposed its industry to create a new society, resulting in a national mindset of exploration that generated new technologies and ways of doing business that were previously unimaginable.[10] We're still benefitting from that change of outlook today. In recent decades, we've seen rapid growth in digitalization and automation in both personal and business domains, making our lives and the ways we work better connected, more productive, and ever more convenient.

Culture Leads Innovation

The postwar rise of consumerism fed off the sense of release following the conclusion of a traumatic period in the world's history. It promised a better life—a good life—to citizens who had experienced war and economic depression over the last two decades.

But as we talk of consumer culture, it's worth taking a moment to examine what is meant by culture itself—and how it continues to impact the way we do business today.

The 19th-century English anthropologist Edward Burnett Tylor opened his book *Primitive Culture* by describing the notion of culture as "that complex whole which includes knowledge, belief, art, morals, law, custom and any other capabilities and habits acquired by man as a member of society." Oxford Languages defines *culture* simply as the "ideas, customs, and social behavior of a particular people or society."

Culture evolves all the time, taking on new ideas and social behaviors, which are intensified and accelerated by influences such as musicians, authors, journalists, artists, and everyday living.

I see culture as the permission slip for business creativity. Entrepreneurs can say that there is a need for their innovations in the marketplace—and they often do, very loudly—but until culture is ready to adopt their innovations, they won't fly. The in-tune entrepreneurs are picking up the vibes from contemporary artists, musicians, and writers as much as from potential consumers. They are waiting for that sudden acceleration when culture gives the green light.

If we look at the dot-com bubble around the turn of the millennium, a lot of brands went too early and went bust as a result. People and the culture they embraced weren't ready for the technology. Ten years later, many of those entrepreneurs and businesses had reinvented themselves with different names and similar products, and society now saw the need.

The Palm Pilot was a good thing. I had one myself. But, ultimately, it failed when the iPhone soared just a few years later. Timing is everything in a world where ever-evolving culture shapes demand.

The environmental space is similar. For years, companies in the nascent environmental sector ticked along, shouting from the rooftops about the need to be green. Now, the public has finally become more sympathetic and responsive, and those far-sighted companies have taken off.

Technology and innovation can come before culture. They can even influence culture. But only once culture catches up will they turn from a good idea into a Force for Good. The trick for business leaders is to be ahead of the curve—but not too far ahead.

The Cycle of Good

Our fast-forward review of industrial history has shown that the pattern of "good brings new" repeats throughout history. We can see this as a Cycle of Good (next page).

The cycle starts with a major change in society. This could be a new cultural movement, an ideological revolution, or a political shift that takes several years—even decades—to become mainstream.

The first movers sense the winds of change. They see what the customers, employees, and communities of tomorrow will need—and they start to innovate. They take the risk of investing in new ideas that might lead to dead ends—or might not. They may even meet with resistance, criticism, and ridicule for challenging perceived wisdom. But they make sure they're in the right place when the right moment arrives.

INFLUENCERS / ACCELERATORS
(MUSIC / JOURNALISM / ART ETC)

CULTURAL
CHANGE

**FORCE
FOR
GOOD**

MAJORITY:
MASS INNOVATION

LIONS:
FIRST MOVERS AND
FASTEST FOLLOWERS

OSTRICHES:
LAGGARDS

STAKEHOLDERS:
CONSUMERS, EMPLOYEES,
COMMUNITIES, INVESTORS,
GOVERNMENT REGULATION

MEDIA

THE CYCLE OF GOOD

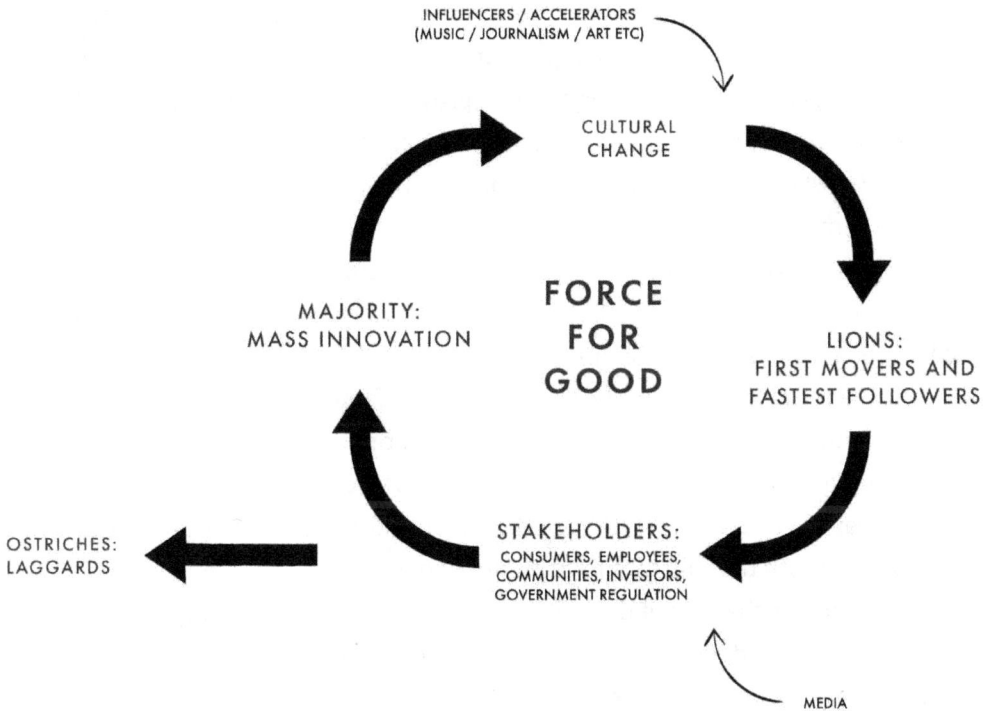

Meanwhile, the incumbents will be largely unaffected and carry on as usual. Some will be ready for change, positioning themselves to become the second movers. Others will be unaware or wish change away. By sticking their heads in the sand, they allow the world to pass them by.

Just to clarify, innovation doesn't have to mean reinvention. Not every business needs to invest heavily in R&D. Innovation is about finding a way to do things differently than before, including "soft measures" such as changing procedures or company culture or bringing in new personnel or partners.

Across different industries, the commonality is that the first movers and fastest followers will get the lion's share of the market. Their instinct is proved correct, and they are duly rewarded. As the vanguard, they gain the prime cuts of publicity and trade. They are also the first to bank public good-will, which can give their brand resilience for the future.

The fastest followers include forward-minded incumbents with the adaptiveness to flex with the changing times. They bring further innovation, as each company finds new ways to meet growing needs. Major brands that have survived and prospered over many decades can bend the zeitgeist to their advantage, whether through acquisitions, partnerships, or their own R&D. Names of inventors such as Gillette, Kellogg, Birdseye, and Biro are still well known today. Over half of the companies in the top 50 most valuable brands are more than 60 years old, from Coca-Cola, Disney, Louis Vuitton, and Chanel to Samsung, AXA, American Express, and Mastercard.[11]

IBM, another stalwart on the list, was nearly a famous laggard itself in the 1990s, having misjudged the emerging market for personal computers.[12] Lou Gerstner, who took over as CEO in 1993, guided the company back to strong health over the next decade. He believed that change and innovation needed to be aligned with the constant pace required for survival.

"Remember that the enduring companies we see are not really companies that have lasted for 100 years," Gerstner said. "They've changed 25 times or five times or four times over that 100 years, and they aren't the same companies as they were. If they hadn't changed, they wouldn't have survived."[13]

As the Cycle of Good continues to turn, the revolutionary idea soon becomes conventional. Challengers are now incumbents; the hunters are the hunted. This is often the juncture at which the government steps in to crystallize the change, writing legislation that benefits the first movers and fastest followers and offering financial incentives to ensure the national market grows. Indeed, the wording of the new rules may have been drafted in consultation with expert advisors from the companies and industries that stand to benefit most. We've now entered the period of mass adoption—until the cycle turns again.

For those laggards who either failed to see the signs, chose to ignore them, or lacked the means of responding, the future looks bleak. Starved of market share and societal goodwill, their eventual demise seems inevitable. Once the regulations are in place, they must join the rest. But even then, it could be too late.

Just as in nature, there are no straight lines in business. No perfect linear progressions. Markets go up and down, hour by hour. Litigation and legislative reform take time, crashing against each other to create unexpected interactions, influences, and outcomes. There are steps forward, followed by backlash and retrenchment, and then the caravan moves on again. The situation on the ground will differ, often greatly, on a local, national, and international level. There are multidimensional, multi-actor, and multiscalar dynamics in play.

We may not always isolate the chicken or the egg. But I believe that, in the long run, business will keep evolving in a positive direction over time, driven by cultural forces. The momentum is too powerful to stop. Those companies that fight against the tide will eventually find themselves swept away.

A Category of One

The brand Toys "R" Us has known how it feels to be both the first-moving innovator and the sinking laggard at different times since its inception in 1946. Founder Charles Lazarus returned from the Second World War with a hunch. Millions of veterans would want to get married, start a home, and have children. He therefore bet on the coming baby boom, taking a $2,000 loan to launch a furniture store called Children's Bargain Town in Washington, D.C. "I would sell cribs, carriages, strollers, highchairs—everything for the baby. My instincts told me the timing was right," he later recalled.[14]

As it happens, his instincts were slightly off. The furniture sold all right, but it wasn't life-changing. Lazarus had the gumption to pivot, realizing that parents might only buy one crib, but they would keep coming back when their children's toys broke or were outgrown. He opened Toys "R" Us in 1957, which sold toys exclusively. After that, his innovative mind turned the business from a single store into a global chain. Flipping the R backward in the logo was clever, as it showed empathy with children learning to write. Iconic toys like Mr. Potato Head found their feet (and ears) at Toys "R" Us. The company mascot, Geoffrey the Giraffe, became a national treasure and rubbed shoulders in promotional events with stars like NBA Hall of Famer Magic Johnson. Lazarus cornered the market by negotiating advantageous contracts, especially with new producers in China, Taiwan, and Japan.

The stores were stacked to the roof, creating an Aladdin's Cave for eye-popping children and their indulgent parents. Lazarus tapped into the feeling of plenty that Americans longed to experience after the austerity of war and the Great Depression.[15] Eventually, his big-box stores sold related goods like diapers and beds, offering a one-stop shop for parents (a smart trick that would eventually be used against Toys "R" Us).

The retailer was also way ahead of its peers in using technology to track and monitor the best-selling products. "I think Toys "R" Us is a unique operation—the only proprietary merchandise company that rivals IBM as revolutionary in concept," a retail analyst told the *Washington Post* in 1982. "Their superb controls and information systems are unrivaled in the industry."[16]

"What we are is a supermarket for toys," reasoned Lazarus in 1981. "We don't have a competitor in variety. There is none."[17] And he was right. The 1,200 Toys "R" Us stores across the U.S. made laggards of many smaller independents, becoming the very definition of a category killer.

In the early 2000s, Geoffrey fronted the new store in New York's Times Square, which boasted its own working Ferris wheel. But the cogs of fortune had already begun to turn against Toys "R" Us. The brand that grew up in the golden age of convenience goods had failed to envision parents' need for even greater convenience. Bigger box stores like Walmart and Target could offer nearly all the bestselling toys as well as everything else that busy parents needed.[18] Kids no longer wanted huge choice; instead they wanted *the* must-have gadget or computer game. The world had changed. Mr. Potato Head was now a movie star, for goodness sake.

E-commerce could have provided a solution for such a well-known brand, but the digital efforts of Toys "R" Us never produced a customer experience to match that of their competitors. A partnership with Amazon burned brightly and then fizzled out. In 2018, Toys "R" Us filed for bankruptcy, leaving thousands of workers redundant.

In 2022, however, Toys "R" Us rose from the ashes, opening 400 outlets in Macy's department stores and growing its global footprint by 50 percent, with 1,400 stores and e-commerce sites across 31 countries. In 2023, the company announced an ambitious "Air, Land, and Sea" strategy to launch new flagship stores in the U.S., as well as offering a "first-of-its-kind" retail experience in airports and aboard cruise ships.[19] Toys "R" Us has become an innovator again.

Faster Cycles, Concentric Circles

Cycles for Good don't spin in a uniform pattern. They whirl off in different directions due to changes in culture, new waves of innovation, and political interventions. Global companies must also adapt to the differing speeds of change in different locations, where local cultures can vary dramatically.

Technology influenced by culture can then influence culture in turn. As I'll discuss in chapter four, the rise of the automobile at the start of the 20th century reflected the growing desire for freedom in society—and then accelerated a vast number of cultural changes, from fast food and vacations to suburbia and sexual liberation.

Keeping up with cultural change is even harder today for businesses compared to previous generations, thanks to the rapid spread of news and opinion on social media and the 24-hour news cycle. Overnight, a business can find itself on the wrong side of a bad news story that will be read, believed, and shared by millions of people worldwide.

The vast amount of noise on the airwaves and in cyberspace also allows cultural movements to form and spread faster than ever. Companies must invest time and effort into keeping up with the pace of change.

Information transparency has increased and will continue to do so, thanks to technology. There was a time when only shareholders had regular access to the inner workings of a business, but now the same intelligence is available to consumers, employees, and society at large, allowing them to make informed decisions accordingly. Of course, that influx of news can work in the good enterprise's favor.

In 2021, a survey by Just Capital found that 90 percent of Americans agreed that there should be a standardized reporting structure for companies to disclose the social and environmental impacts of their business practices. The same number agreed that "the activities and behaviors of America's largest companies impact society as a whole." Even on climate, a subject that typically receives a partisan response, nearly 90 percent were in favor of mandatory disclosure of environmental impacts.[20] That's a cultural movement, right there.

As global trends, whether around climate change or inequality, continue to impact society and stakeholders become better informed, enterprises will recognize a sustainability-led approach as more than the right thing to do—it will be a financial imperative.

The Benefits of Fast Culture?

Society is demanding ever-greater transparency among companies—and today's always-on media brings benefits in that respect. Rewind to 19th-century Britain, where working conditions in factories were constantly criticized by contemporary campaigners. Children were employed from a young age to reach underneath moving machinery or climb into narrow mine shafts, with inevitable tragedies the result. Women were expected to fulfill a long list of domestic chores and work 12-hour shifts outside the home.

Yet despite public concern, the pace of reform was extremely slow. Might public pressure through social media channels have led to faster change, had those channels been available? Modern labor laws putting pressure on companies to make their labor practices more humane might have come sooner.

The rise of newspapers in the 19th century, and then cinema, radio, and television in the first half of the 20th century, directed more attention to business practices, making their owners and shareholders ever-more answerable for corporate scandal, environmental and social disasters, and poor working conditions. Yet it should be noted that numerous multinational companies and their CEOs were able to handle major incidents with relative ease, at least until recent decades.

The BP Deepwater Horizon disaster off the Coast of Mexico in 2010, in which an oil rig explosion killed 11 people and released the largest marine oil spill in U.S. history, became a global news story within an hour. BP's CEO resigned, the company's stock price fell by 51 percent, and the company had to set up a multibillion-dollar fund to pay for damages.

CEOs now understand the need to keep their businesses one step ahead of damaging headlines. Many encourage whistleblowing to identify potential problems in safety and risk management before they escalate.

An individual business may need to follow several different cultural movements relating to issues such as shifting work patterns, diversity, and

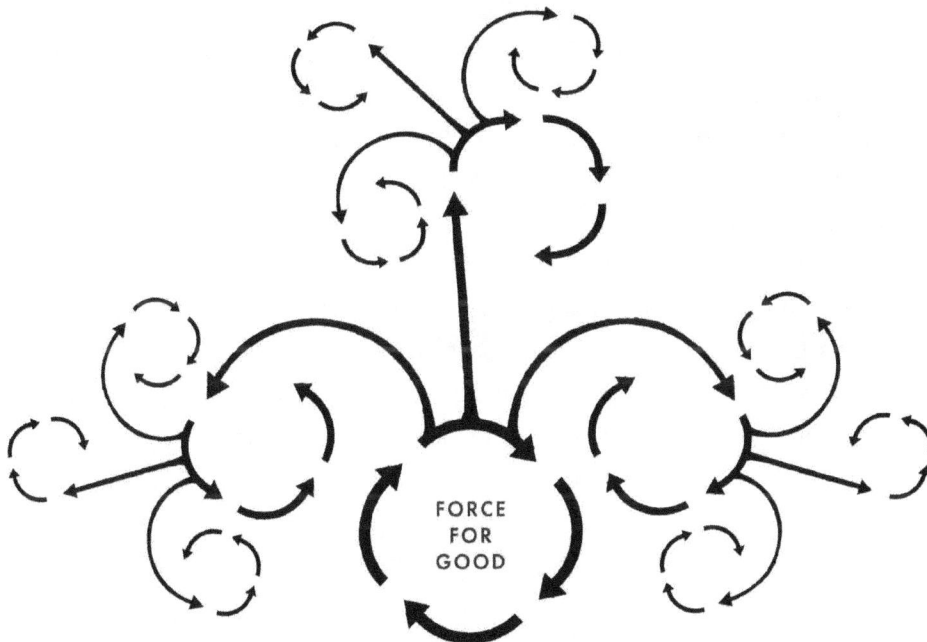

A SPIROGRAPH OF CYCLES

environmental responsibility. The result is a pattern of concentric circles, creating a "Spirograph of cycles," each requiring an innovative response (previous page).

Moreover, the cycles are spinning faster, leading to shorter time frames for innovation. The adjustments are harder to find, but staying up to speed with cultural change is critical.

Business Is the Agent

What's important to note here is that lasting change in society is often in the hands of business. Cultural movements or even political pressure may provide the spark, but business is the machinery of change. Through innovation, it's business that creates a Force for Good.

Smart business leaders envision their future from a cultural perspective. They ask, "Where's the world going?" Culture is driving market opportunity, and they are getting on board. In fact, for society to move forward, it needs business on board.

Furthermore, doing good once isn't enough. Businesses must repeatedly be a Force for Good to survive and grow sustainably. The cycle also shows that profit-driven organizations and individuals can be a Force for Good in society by following their natural motivations.

The definition of good changes over time. How we perceive good now would have been deemed absurd in the past, while certain practices among our ancestors in the name of good are abhorrent today. Will future generations look back on us with envy, curiosity, disbelief, or even disdain?

Finding Comfort in the Past

Sustainability issues just happen to be the defining problems of our time. Eighty years ago, only a few people would have felt safe in forecasting climate change or the progress of artificial intelligence via machine learning. Eighty years from now, the challenge of the age could be something beyond the imagination of contemporary sci-fi writers.

Yet there's nothing to suggest that the pattern of cultural change leading to innovation and advancement won't hold true as it has throughout history. Culture is the catalyst that moves businesses to become a Force for Good. And the necessary innovation typically comes long before regulation. When it comes to environmental action, as an example, good businesses don't just shrug and say, "Okay, we'll change when we have to." They react by asking, "What innovation can we put in place to ensure our production processes are sustainable?" They come up with safer machines for the workplace, smokestacks that emit less, carbon capture methods—whatever is needed. They also find ways to measure and broadcast their progress, whether it takes the form of reduced carbon emissions or a more diverse board. They turn challenges into opportunities to stand out.

I believe that this cycle should make us feel positive about the future. History shows that, as challenging cultural pressures and political trends appear, business innovation finds a way. You can draw a line between what society needs and the major technological developments of the next 10 years.

The message is that culture affects business. And then business wakes up and does what it does best, which is innovating and being a Force for Good. We can see that through looking at the past, and this provides a positive message for us in the future. We're not doomed, after all.

There is a temptation to look back on the past with nostalgia, yet so much of our world is better than it was even 20 years ago, let alone 150 years ago. Of course, there are exceptions, but levels of health and safety at work, for example, have vastly improved. Industrial pollution, such as catastrophic chemical spills, could have flown below the radar in the past, but businesses and their leaders face greater scrutiny today. Companies feel increasing pressure from their investors to ensure that their supply chains abroad meet the same levels of compliance.

Innovation drives prosperity. Some businesses may look at a problem and scratch their heads, thinking that it is just plain unsolvable. But then one company finds a solution, which quickly becomes standard. This break-through creates new jobs and even new sectors, sparking fresh innovations and wider prosperity for people in general. History shows this sequence time and again. Those Force-for-Good companies do best in terms of employee retention, community impact, and, yes, even shareholder value.

Value to and from the Stakeholder

Another lesson from history is that innovative people and businesses recognize the need for balanced actions that work within the world they're living in. Too much give or take becomes counterproductive. Others step in to redress the balance, at the expense of those that have upset it.

For today's businesses, we can see this necessary equilibrium as a tradeoff between value to stakeholders (VTS) and value from stakeholders (VFS)—whether those stakeholders are consumers, employees, communities, governments, or investors.

Value is continually being traded among all these stakeholders. Obviously, consumers provide revenue when they buy a company's goods. Employees provide their labor, communities provide the license to operate, and governments provide regulations that enable fair trade. Businesses offer various forms of value in return: high-quality products to consumers, fair wages and benefits to employees, good-neighbor policies in relation to communities, and tax payments to governments. Any imbalance can lead to a breakdown in the relationship, and then to suboptimal business performance (next page).

There's a tendency to talk about maximizing shareholder value and profits. In the near term, businesses can maximize returns by producing a terrible product, polluting the environment, violating regulations, and treating their employees badly. They may gain a tremendous amount of revenue, but it won't last. Customers will leave. Employees will quit. Their reputation will tank. The community will shut them down for abusing local trust. The government will come calling. Very soon, they'll find themselves out of business

Appropriately balancing the value derived from the stakeholder and the value
provided to the stakeholder results in an Optimal Stakeholder Relationship

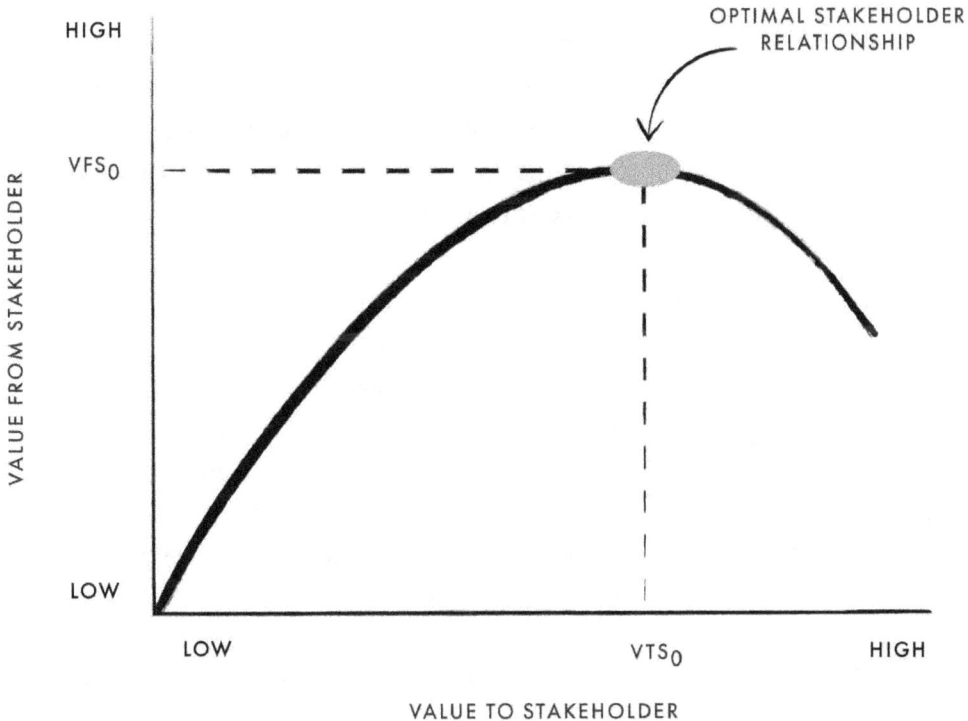

OPTIMAL STAKEHOLDER
RELATIONSHIP

HIGH

VFS_0

VALUE FROM STAKEHOLDER

LOW

LOW

VTS_0

HIGH

VALUE TO STAKEHOLDER

THE OPTIMAL STAKEHOLDER RELATIONSHIP

and maybe in court. Their VTS / VFS equation is wholly imbalanced towards
VFS.

It can go the other way, where VTS dwarfs VFS. A community may sting a
local enterprise with higher taxes to the extent that the business feels forced
to leave the jurisdiction, taking jobs with it. Regulations can be too punitive,
and the whole region or country loses out to competitors. If a company provides customers with goods and services at a price that impacts profitability,

then it will ultimately fail. Paying employees twice the market rate may attract top talent, but is it feasible in the long term? Without a sense of balance, it becomes impossible to run a profitable business.

A company might follow a herd mentality by providing various features in its offering that its competition is also providing, with little understanding of the cost implications of the additional features. Financial service firms, for example, frequently provide products and services below cost, with little understanding of whether such services provide sufficient incremental value to the customer to justify decreased margins in terms of value from the customer. Thus, profitability is at risk when the customer and the organization perceive VTS differently.

Financial institutions also got burned when they paid traders exorbitant bonuses to take risks, only to find that rogue decisions ended up bankrupting the whole business. Now, at smart companies, traders are paid over a longer time frame to incentivize a longer-term perspective. Likewise, CEO equity compensation is paid on a multiyear basis. Many companies are practicing profit sharing above the hourly wage to incentivize workers to be on time, do a good job, and maintain a clean environment. In this way, employees are encouraged to hold a stake in the business.

Good sustainability strategies should help businesses find an equal exchange of stakeholder value. They will involve all stakeholders, helping to keep the dial pointed at optimal.

The most important time to consider this balance is at the point of innovation. When you build a product and start employing people, or figure out where to put your plant, you need to be asking, "Where does the balance lie? How can I optimize value to and value from my stakeholders?" Again, that's where enterprise sustainability comes in, because the impacts on people and the planet give you all that information. If you don't get this right at the beginning, then you're not getting the benefits from your stakeholders from the outset.

In particular, examine the proposed innovation from a customer-centric perspective. Is there enough value coming back to you? Can you build the business innovation at an economically competitive price? Will there be a sufficiently large market? Can you make a profit? Will you get capital?

Getting the answers to these questions is a challenge in itself. It requires you to find the right people and the best technology. It also requires you to follow the data. By aggregating the necessary information, you can then analyze it to find out where your stakeholders are today. This baseline will tell you how government regulations, the going rate of pay, environmental risks, and other factors will impact your VTF / VTS equation.

You also need to monitor your innovative efforts as they unfold. If a new initiative didn't work out, why not? Did you price it right? Did you have the right level of talent?

With these kinds of information in hand, you can design and prioritize your initiatives to ensure the appropriate allocation of capital to those most likely to increase stakeholder value and profitability for your business.

Not all businesses are created equal, and neither is every sustainability strategy. A financial institution, a heavy manufacturer, and a retail business will have very different stakeholders. So it's important to optimize on a customized basis.

Try viewing your next strategic step through the filters of VTS and VFS. Your strategy and decisions may well differ from those of other companies. If you are running a Force-for-Good, sustainable business, you may find that most of your core decisions stay the same. A lot of successful CEOs consider their stakeholders instinctively. Good businesses have been doing this forever. For them, it's inconceivable not to take stakeholders into account.

Ultimately, the key is to balance the value provided with the value received. Get that right, and the results will be good. Perhaps even very good.

In the next chapter, I'll dive deeper into how a successful business is set up as a Force for Good—and how its leaders balance their VTS and VFS.

Key Takeaways

- Cultural change has proved to be a catalyst to transform businesses toward good throughout history.
- People have long understood the concept of good, and they create movements to promote it when they find it to be absent.
- The entrepreneur in a capitalist society must learn to recognize cultural shifts before others in order to innovate in a way that promotes the good.
- Those who take the risk to further the good will be ahead of the pack and benefit accordingly.
- In today's society, Cycles for Good are spinning faster, demanding ever-greater agility, transparency, and vigilance.
- Sustainable businesses understand the need for constant change and for a careful balance between the value to and from their stakeholder.

3

THE FORMULA FOR GOOD

- Good business combines expertise, technology, and data
- This formula stretches back over centuries . . .
- . . . and endures today in every sector
- How can businesses find the right people and keep them?
- How can we back the right technology—and the next technology?
- How can we get more from rapid advances in data?

"Information is the resolution of uncertainty."
—Claude Shannon, inventor of the digital bit

In the first two chapters, I've looked at the importance of being a Force for Good as a business, and how the Cycle of Good, linking culture to business innovation, acts as a means of powering society forward. In this chapter, I'd like to shift focus from the *why* to the *how* of good. Is there a methodology that can help businesses read the cultural tea leaves—and then convert those intuitions into competitive innovation?

I believe that the combination of expertise, technology, and data is an essential mix for companies seeking to become (or remain) a Force for Good. By harnessing this combination, businesses can secure the early-mover advantages of innovation and strengthen their organization for long-term growth. In doing so, they can meet the needs of shareholders and stakeholders alike.

These three factors should be guided by a Noble Purpose, without which companies can quickly lose focus, precision, and orientation. Expertise, technology, and data can prove a force multiplier for that Noble Purpose—an essential foundation for transforming a vision into reality.

I see this relationship as a Formula for Good:

NOBLE PURPOSE × (EXPERTISE + TECHNOLOGY + DATA) = GOOD

Expertise refers to working with experts in their fields, welcoming their knowledge and experience as an accelerator of change in the business. Good minds are a Force for Good.

Technology means constant vigilance around the right technology—and the next technology—that will catalyze your business performance. This includes both the adoption and the invention of necessary technology.

Data is increasingly the lifeblood of a successful business, providing the signals and stimuli that will identify new opportunities and then guide your response. Data tells you where you've been, where you really are right now, and where you need to get to in the near and, hopefully, in the long-term future.

When multiplied by a Noble Purpose, these three forces can help businesses create an almost limitless amount of good, benefiting both themselves and all of their stakeholders.

A Modern Medical Miracle

A recent example of the Formula for Good was the race to develop a mass vaccine for the Covid-19 pandemic. In 2020, the disease claimed the lives of an estimated three million people and caused untold disruption to communities and businesses worldwide.[1] The International Monetary Fund estimated the total cost of the pandemic to the global economy through 2024 at $12.5 trillion.[2] The development of a vaccine appeared to offer the quickest escape from the cycle of lockdowns, fear, death, and economic uncertainty.

But vaccines typically take many years, if not decades, to develop. (The previous record was four years, set in the 1960s for tackling mumps.[3]) And finding the vaccine wouldn't help without efficient rollout to millions of people in a short time frame. This was both a logistical and a medical challenge. Any delay would cost more lives and cause more economic harm.

Moreover, developing the vaccine *during* a worsening pandemic of a novel coronavirus presented a unique conundrum for biopharma. How could they demonstrate that a vaccine was indeed effective? And once the green light was given, how to avoid a debilitating lull in manufacturing the doses?

Fortunately, expertise, technology, and data were deployed right at the heart of this Force for Good.

The vaccine developers were not starting from scratch. They sought to create a new type of vaccine that uses a molecule called messenger RNA (mRNA) rather than part of an actual bacteria or virus. The technology behind mRNA vaccines had been studied before Covid-19, having been trialed on the flu, the Zika virus, and rabies, in addition to cancer research.[4] The pioneers at BioNTech in Germany, who teamed up with U.S. pharma giant Pfizer, had battled for many years to bring their theories regarding mRNA vaccines to a wider audience.[5] Like many successful innovations, it was waiting for the moment in which it was needed the most.

Still, the development of the Covid-19 vaccine required unprecedented collaboration between industry and academia to ensure that the manufacturing processes could meet demand. The AstraZeneca-Oxford University vaccine carried out multiple manufacturing and testing processes "in parallel instead of consecutively—saving months of time compared to a traditional vaccine manufacturing process."[6]

Data analytics helped to accelerate process development, supporting robust scale-up and tech-transfer methods and improving manufacturing processes, according to the life sciences company Sartorius.[7] Data visualization was critical for informing the public about the spread of the disease, broadcasting the urgency to isolate and the availability of vaccines. Covid-19 was the first disease in history that you could track on social media.

The traditional means of communication in the medical community—submitting papers and attending annual conferences—would have extended the process by months, if not years. Now ideas about controlling the pandemic were instantly posted and digested by experts around the world, who could react and give feedback almost immediately. The speed of information flow through technology proved a game-changer. Constant data monitoring is now needed to stay ahead of new variants with adapted vaccines.

The role of medical experts such as experienced virologists, epidemiologists, and public health practitioners was particularly important for maintaining calm and conveying critical information. In the race to be the first to introduce an effective vaccine, pharmaceutical companies had to rapidly recruit the right teams—not an easy thing to do during a public lockdown. The almost overnight transformation of working practices from in-person to virtual was another technological miracle of the pandemic.

By December 2020, within just nine months of the World Health Organization declaring Covid-19 a global pandemic, the first members of the public were receiving doses of a vaccine. More than 5.55 billion people worldwide have since received a Covid-19 jab, equal to about 72.3 percent of the world population, according to the *New York Times* vaccinations tracker.[8] An estimated 20 million deaths were averted by the vaccines.[9] Without those lives, the global economy would have been almost $9.2 trillion worse off.[10]

Once the rollout of vaccines was underway, accurate data was needed to prove that people who might have died were now surviving. This information would directly save lives. Scotland produced the first real-world data on single-dose vaccine effectiveness just two months after the rollout began in the

U.K. A national data infrastructure "linked vaccination, primary care, COVID-19 testing, hospitalization and mortality records for 99% of the Scottish population."[11]

The results indicated a high level of protection after a single dose—even for the elderly—which increased public confidence in the U.K. vaccine rollout. The ripples were felt further afield, as the research influenced the national regulatory strategies in Canada, Denmark, France, and Germany. Data was also gathered in the U.K. to help understand which sections of society were hesitant about accessing the vaccine.[12]

More widely, the pharmaceutical industry illustrates the benefit of the Formula for Good in their drug research. AstraZeneca, for example, brings together experts from all over the world into its three lead R&D centers in Gothenburg (Sweden), Cambridge (U.K.), and Gaithersburg (U.S.). Data and technology are critical for designing new treatments for cancer, cardiovascular diseases, and respiratory diseases.

Understanding the Value of Green

The sustainability sector is another area that's actively bringing the Formula for Good to life. The green technology and sustainability market worldwide is projected to reach $74.64 billion by 2030, growing at a compound annual growth rate (CAGR) of more than 20 percent by the end of the decade.[13] Presented with this multibillion-dollar opportunity, businesses and investors are switching on to the long-term opportunities provided by industry innovation and government funding.

Although the environmental movement has a long history (see chapter five), the sector is still taking shape and growing fast. Established experts are now in high demand. And the proliferation of sustainability schools in leading universities, themselves backed by billion-dollar budgets, points to a rapid injection of even more talent in the coming years. Startups and entrepreneurs in green technologies such as renewable energy generation, electric vehicles, energy storage, carbon capture, and household energy-saving devices—to name but a few—are attracting major investors who can see these innovations shaping the future of society.

Data plays an increasing role in defining the relationships among businesses and all of their stakeholders. For example, data provides evidence for unlocking investment. Sustainability metrics around environmental and social concerns provide companies with the license to operate. Regular data reporting on sustainability issues such as carbon emissions will become table stakes for companies. Beyond life-cycle analysis of their own operations, businesses on the path to net zero will need to present a holistic picture to all stakeholders, including the emissions generated throughout their supply chain (a form of analysis often referred to as the Scope 3 challenge). In due course, standardization of data collection and calculation will facilitate more accurate and transparent reporting, in addition to data-sharing both within and between organizations.

Revolution Is in the Air

Looking at history, we can now see how the Formula for Good has guided the four industrial revolutions of the last 250 years. The First Industrial Revolution was known for innovations in water, coal, and steam power. The Second, sometimes called the American Industrial Revolution, harnessed steel and electricity. The Third, which dominated the last third of the 20th century, saw the rise of electronics, information technology, and automated production. The Fourth Industrial Revolution is now capturing the potential of smart automation, artificial intelligence, big data, and interconnectivity.

Let's take a closer look at how the combination of data, technology, and expertise helped to shape each of these major steps forward.

The steam engine is one of the enduring symbols of the First Industrial Revolution, which propelled the economic transformation of Britain during the 18th century. As the major coal producer of the time, Britain had a particular need to solve the problem of flooding in mines, which slowed productivity and presented a severe risk to miners. This pressing demand provided a commercial incentive for innovation, pitching the leading engineers of the day into a battle to develop an effective steam engine that could pump the water out.

In 1784, Scottish engineer James Watt won the top prize by inventing the double-acting engine, in which the piston pushed as well as pulled. This required Watt's development of so-called parallel motion, an arrangement of connected rods that guided the piston rod in a perpendicular motion, which he described as "one of the most ingenious, simple pieces of mechanism I have contrived."[14] More efficient and cost effective than its rivals, Watt's steam engine accelerated the mechanization of factories, such as cotton mills.

Although it is Watt's name that has passed into history, his achievements owed much to those who preceded him and to the moment in which he lived. His impact was immense, given the scale of societal change that followed the Industrial Revolution, but it's not fanciful to suppose that somebody else could have reached the same conclusions quickly enough, such was the hunger for innovation at the time. A large factor in Watt's success was the open-source approach to data. Knowledge was actively diffused among the engineering community, both at home and abroad, allowing engineers to learn from one another.[15]

Watt lived in an age when data was beginning to dominate science and technology. The modern science of statistics was then in its infancy. For the first time, tables were being used for recording everything from temperatures, rainfall, and agricultural inputs and yields to the hardness and softness of materials. Watt himself employed extensive tables, graphs, and mathematical models to deduce power and fuel efficiency. Watt's career exemplifies how data, technology, and expertise worked together to make the First Industrial Revolution possible.

The Man Who Listened

If water powered the First Industrial Revolution, then steel built the Second Industrial Revolution. From railroads to shipping, bridges to skyscrapers, the sudden availability of affordable, incredibly durable steel allowed industrial-ists to modernize the U.S. at a remarkable pace in the second half of the 19th century.

Steel magnate Andrew Carnegie was one of the primary drivers of this transformation, amassing a business empire from the 1870s to the early 1900s that would eventually sell for $480 million (close to $13 billion today).[16] He was another of those prominent personalities of history whose fortune was greatly aided by a combination of expertise, technology, and data.

Born into a poor weaving family in Scotland in 1835, young Andrew soon emigrated to Pennsylvania, where he found work as a bobbin boy in a cotton factory. But he was ambitious and tapped into the sense of possibility that permeated the U.S. at that time. Carnegie educated himself at night school, which gained him a job as a telegraph clerk at the age of 14. Spotted by the superintendent of the Pennsylvania Railroad Company, Carnegie was headhunted for the role of private secretary and personal telegrapher before rising to superintendent when the vacancy arose. He showed a shrewd nose for investment opportunities, growing his annual income to $50,000 (nearly $1.5 million today) by his thirtieth birthday.[17]

On one of his regular visits back to Britain, Carnegie witnessed the new Bessemer process, named after its inventor Henry Bessemer, which solved the challenge of manufacturing steel that was affordable *and* free of impurities. Carnegie immediately recognized the potential of bringing this technique to the U.S., and set about introducing Bessemer converters in Pittsburgh. By 1889, the Carnegie Steel Company was dominating the industry.

Carnegie was more than merely the right man at the right time. He was rigorous in his use of data to identify efficiencies that reduced the cost of pro-duction. Carnegie was also one of the first to introduce the industrial strategy known as vertical integration, buying up coke fields and iron-ore deposits in addition to the shipping and railroads needed to transport his steel.[18]

He also had a good eye for expertise, surrounding himself with capable lieutenants such as the engineer and manager Charles Schwab, who was one of the first Americans to earn a million-dollar salary due to his unique ability to "arouse enthusiasm in his people." Carnegie had a vast executive team, as he wanted specialists to report on every aspect of his business. "The test of any man's ability is not what he does himself, but what he can get others to do in co-operation with him," he told journalists in 1899. "No man will make a great leader who wants to do it all himself or get all the credit for doing it."[19]

A "harmonious spirit" was a non-negotiable for Carnegie when recruiting people, according to Napoleon Hill's 1937 global bestseller *Think and Grow Rich*, which he wrote on Carnegie's recommendation.

"As I grow older, I pay less attention to what men say. I just watch what they do," was another of Carnegie's dicta. He was a great observer, challenging his staff to show him their skills, rather than tell him in reports. Once satisfied, he was ready to give his people responsibility and invite challenge in return, as the situation arose.

It was Schwab's idea to sell the business to financier J.P. Morgan's newly formed United States Steel Corporation in 1901, forcing Carnegie into retirement. He subsequently dedicated himself to philanthropy, declaring that great accumulators of wealth should bestow any surplus to "the improvement of mankind" through charitable causes. Some $350 million was endowed to causes and venues in the U.S. and his native Scotland, with a focus on supporting education and research in law, economics, medicine, and international peace. Carnegie is remembered as a business magnate who harnessed expertise, technology, and data to help transform the United States into the world's leading industrial power by the start of the 20th century.

The Information Age and Beyond

Since the end of the Second World War, expertise, technology, and data have been applied to underpin some of the greatest societal advances of the Third and Fourth Industrial Revolutions.

In the late 1940s, Bell Labs, the dedicated R&D department of AT&T, released several innovations that fundamentally changed the way the modern world works. For example, the transistor—the small semiconductor device invented by Walter Brattain, John Bardeen, and William Shockley—is at the heart of all modern electronic technology, from telecommunications and data communications to aviation, space exploration, and audio and video recording equipment. Their discovery owed much to the recruitment policy of Bell Labs, which invested heavily in bringing the very brightest talent to its open-plan laboratories, where they were encouraged to mingle and share ideas. Brattain was a theorist, Bardeen a materials guru, and Shockley an experimentalist.[20] They were left to their own devices, so to speak. The three colleagues scooped the Nobel Prize for Physics—one of nine Nobel prizes won by Bell Labs.

The following year, in the *Bell System Technical Journal,* Claude Shannon published his seminal work, "A Mathematical Theory of Communication." In this article, he outlined his information theory, which would become the cornerstone of the modern digital information age, showing how logical statements could be translated into 1s and 0s.[21] Today, the binary digit (bit) is still the universal currency of data.[22] Dr. Robert G. Gallagher, an esteemed professor of electrical engineering, said of his colleague at MIT, "Shannon was the person who saw that the binary digit was the fundamental element in all of communication. That was really his discovery, and from it the whole communications revolution has sprung."[23]

Shannon felt particularly at home in the Bell laboratories, riding a unicycle while juggling in the corridors or playing with his mechanical mouse Theseus, which he tried to navigate through a maze using early forms of artificial intelligence. "I visualize a time when we will be to robots what dogs are to humans. And I am rooting for the machines," he wrote. "My mind wanders around, and I conceive of different things day and night. Like a science fiction writer, I'm thinking, what if it were like this? I am very seldom interested in applications. I am more interested in the elegance of a problem. Is it a good problem, an interesting problem?"[24]

Scientists at Bell Labs developed plenty of other elegant solutions to complex problems, including the laser, the photovoltaic cell, the first orbiting communications satellite (Telstar 1), the Unix operating system, and several programming languages.

In another field, around the same time that Shannon launched the bit, what became known as the Green Revolution was also demonstrating the power of expertise, technology, and data in greatly increasing food output in countries with severe food poverty.

In 1943, American agronomist Norman Borlaug began work on a project backed by the Rockefeller Foundation in partnership with the Mexican government to tackle the country's ailing corn production. By analyzing the behavior of over 800 varieties of Mexican corn under diverse conditions, the team used the data to identify the most suitable varieties. They then worked closely with local farmers, and crop yields improved almost immediately.[25] Borlaug and his team continued and expanded the work, combining technologies such as synthetic fertilizers and scientific plant breeding in Mexico, Africa, and Asia. Today, the Green Revolution they launched is credited with saving the lives of more than a billion people from starvation.

The Rockefeller Foundation is currently supporting other initiatives that use data and technology to "unlock a parallel revolution" in food and agriculture. For example, the tech startup WeRobotics flies drones in rural agricultural communities to analyze soil, monitor and assess crop health, and assist with planting and spraying—with an emphasis on empowering local experts.

Innovation Is All Around Us

It's not hard to see other current examples of progress fired by expertise, technology, and data in everyday life. Think of how entertainment has opened up through digital streaming channels such as Netflix and Spotify, curated by data to bring greater choice to global audiences. Just as artists from Bach, Mozart, and Beethoven to Elvis Presley, Aretha Franklin, Paul McCartney, and David Bowie influenced music in their lifetimes and beyond, AI composition and the democratization of sampling is now producing whole new kinds of electronic sounds that are gaining worldwide audiences. Big data analytics allows the music industry to tap into the cultural zeitgeist to predict the next big hit. In retail, brand managers increasingly use data to measure and analyze the individual customer journey and buying choices to personalize every aspect of the experience and second-guess the next purchase.

Sports such as Formula One are a test of each team's technological innovation and real-time data analysis, in addition to the ability of the race car drivers. The steering wheels alone cost thousands of dollars and are essentially a game console for drivers, who must react to data streams in a split second—with their lives in the balance. Manufacturers can test drive their own R&D in terms of performance, fuel economy, and safety, which will ultimately end up in the cars we all drive on the highway.

Professional sports franchises use new technologies to monitor the real-time fitness, nutrition, sleep patterns, and recovery times of their players while scrutinizing the strengths and weaknesses of opponents. A marginal gain might prove the difference between success and failure, profit and loss. An incoming coach or star playmaker could raise the mood of millions around the world. Major sporting events like the Olympic Games, FIFA World Cup, and the NFL Super Bowl unite industries as varied as telecommunications, marketing, retail, food and beverages, transport, sustainability, gaming, education—you name it—all fueled by expertise, technology, and data.

There are many uplifting examples of how the Formula for Good has changed lives in modern society. Thanks to digital technology, people with disabilities have greater opportunities to participate fully in society, stay in touch with loved ones, access education, and appreciate music and entertainment.

Louis Braille's invention of a new tangible code for visually impaired people was one of the great technological advances of the 19th century. His work has opened many doors in terms of communication, knowledge and imagination. Today, digital braille displays allow blind people to read about current affairs, consume books and respond to messages on the move. Similarly, digital implants have improved the lives of those with hearing impairments, while a wide variety of apps such as live transcribing can help users keep up with group conversations. Using other apps that harness musical vibrations, the hearing impaired can enjoy playing or experiencing music at home or at concerts.

In business, the speed and ease of collaboration and knowledge-sharing have greatly increased as a result of the technologies that were refined during the Covid-19 pandemic. On average, businesses that use web conferencing save 30 percent in travel expenses, which is good news for the financial, social, and environmental bottom lines.[26] If only we could measure the commercial benefits that come from the rapid transfer of new ideas through online meetings and social media.

Taking Tech to the Next Frontier

If we look at each element within the Formula for Good, we can see their continued impact on business today. Let's focus on technology, for example. In 2022, research by Accenture found that "market leaders who are stepping up their investments in cloud, AI and other technologies are actually growing their revenues at up to five times the rate of laggards." While leaders in tech adoption made up the top 10 percent of the study and laggards the bottom 25 percent, Accenture also identified a band of "leapfroggers" among the rest, who were gaining fast on the leaders due to their foresighted investments in new technologies.[27] These on-the-move companies are "building core systems strength and scaling new technologies. They also flip their IT budget to favor innovation" over "keeping the lights on." In contrast to the 30 percent share usually spent on innovation and the 70 percent spent on maintenance and operations, they reverse that split.

According to the research, leapfroggers outspend their peers in cloud innovation and grow their tech adoption by an average of 17 percent, while nearly all are building partnerships across their ecosystems. They are expanding "access to technology across all functions" and embracing tech as a way to "address employee reskilling, well-being and mental health."

Leapfroggers typically implement a top-down strategy that sits on the radar of every member of the C-suite, not just the CIO. They use diagnostics to benchmark their results from technology. They also evolve their culture to improve agility, expediting their progress to viable products. Talent transformation is key to diffusing tech knowledge throughout the organization, backed by a cloud-based technology foundation.

Every company can learn from the tactics of the leapfroggers. Not every business will need to invest in emerging technology such as blockchain, augmented reality, 3D printing, or robotics. However, applications such as deep learning, machine learning, Internet of Things (IoT), edge computing, data lakes (in which large amounts of raw data are stored), and big data analysis will only increase in importance—indeed, many day-to-day processes rely on them already.

Winning the Talent Game

Just as a star quarterback or virtuoso violinist is valuable in sports and music, so the ability to hire and retain qualified tech workers may prove the difference between success and failure for companies. The shortage of trained candidates for these positions is a concern for over half of global tech leaders, according to a poll by MIT Technology Review Insights.[28]

Data scientists, software engineers, programmers, and cloud computing experts are at the top of the wish list for companies. And it's not just the usual software and tech powerhouses in search of talent. Retail brands, banks, insurance firms, and even schools and government departments are increasingly dependent on tech expertise. Recognizing these realities, many countries are investing in education in technology to build their workforces.

More than two-thirds of global businesses will embrace emerging technologies in areas such as Web3, the metaverse, and edge and quantum computing within the next two years, according to KPMG. However, ongoing talent shortages could complicate the adoption of these technologies.[29] More than half of respondents said they were behind schedule on cybersecurity due to the talent shortage. The so-called Great Resignation experienced by many companies following the Covid-19 pandemic has exacerbated these problems.

Expertise, then, is a valuable commodity, one that can be hard to find in a competitive talent market. The global consulting industry, which is expected to grow by almost eight percent annually to $1.3 trillion in 2026, is poised to help fill the shortfall.[30] This rise is due in part to the ability of consultancies to process and analyze massive amounts of data, providing accurate insights into key areas such as operations, HR, sales, procurement, and supply chain activity. The need to gather, process, and report sustainability data will likewise increase the demand for reliable expertise.

In this competitive environment, companies are seeking new approaches to improve their talent acquisition as a means of survival. Successful tactics include strengthening the workplace culture and making tighter links with universities through internships. There's also a realization that it pays to develop individual potential. By investing in candidates who don't hit the same high notes in exams but demonstrate qualities such as loyalty, humility, hard work, and readiness to learn, companies can train them in the right skills later.

The ongoing war for talent is having an international impact. In recent decades, the global ratio of high-skilled migrants to low-skilled migrants has ballooned in response to the growing need. Countries that offer the most attractive opportunities are reaping the benefits. Nearly 75 percent of all high-skilled migrants live in the United States, the U.K., Canada, and Australia.[31] Over 70 percent of software engineers in Silicon Valley were born abroad.[32] Many of the world's most famous brands have been led by immigrant CEOs, including Google, Microsoft, Coca-Cola, Pepsi, Pfizer, and Honeywell.

Smart companies recognize the value of today's global talent pool. According to research by the Boston Consulting Group, the most diverse companies are the most innovative. It found that those companies with "above-average total diversity had both 19% higher innovation revenues and 9% higher EBIT margins, on average."[33]

Global talent mobility can also provide an upside for source countries, as their brightest minds are exposed to world markets, creating a diaspora effect that feeds innovation back home.[34] For example, the fast-growing tech communities in India and China have provided a steady stream of talent to the West—and many of those recipient Western companies have now set up offshore hubs in India and China. Technology is also powering virtual talent mobility with video conferencing, online labor exchanges, and digital platforms.

Going Further with Data

Of the three factors in the Formula for Good, data feels the most "modern." As noted, metrics and analysis have helped to build civilization since records began, but the word *data* itself (from the Latin for "a fact given") only arrived in English in the 1640s.

The rise of capitalism in the early modern period coincided with a growing appreciation for the value of data. In the 1660s, a business owner named John Graunt studied death records in London parishes to make observations on gender-related death rates, cause of death, and even life expectancy. For a penny, Londoners could buy his early big data analyses to make decisions on the best place to live.

Today, data visibility is key to almost any business. As a result, data visualization has become a valuable skill for presenting stories to investors, employees, the media, and the public alike. It's an art form that has been practiced with increasing sophistication for almost two centuries. One classic example: In 1869, the French cartographer Charles Minard drew a band graph that poignantly captured the disastrous Russian campaign conducted by Napoleon in 1812–1813. Minard plotted the French Army's successive fatalities on its journey to Moscow and its subsequent retreat. The vivid image he created—complex yet surprisingly easy to interpret—has been described by data visualization guru Edward Tufte as "the best statistical graphic ever drawn" (next page).[35]

Technology has played a vital role in our growing reliance on data. The 1880s saw the first mechanized example of data processing in response to the U.S. census, which was creating more data than collectors could reasonably analyze. Herman Hollerith, an inventor who worked for the U.S. Census Bureau, created a machine that used punch cards to process the data far more quickly than older manual methods. Hollerith launched his own company, which would later amalgamate with others to form IBM.

In 1928, Fritz Pfleumer worked out how to record data on magnetic tape, paving the way for hard disk drives and floppies. In 1958, IBM researcher Hans Peter Luhn defined business intelligence as "the ability to apprehend the

MAPPING NAPOLEON'S MARCH, CHARLES MINARD, 1861

interrelationships of presented facts in such a way as to guide action towards a desired goal."[36] It's still hard to top, as definitions go.

In the 1960s, Dr. Joseph Carl Robnett Licklider envisioned the concept of cloud storage with his proposal for an Intergalactic Computer Network, although it would take several decades for cloud computing to become a working technology. In the 1970s, IBM's Edgar F. Codd coined the term "relational database" (think Excel spreadsheets with columns and rows), which is still used today as a framework for data management.

In the 1990s, Tim Berners Lee created hyperlinks and hypertext for the internet, facilitating the sharing of data on the World Wide Web. Google Search soon followed, putting data at the fingertips of anyone with a web connection. Global Positioning Satellites (GPS) began pinpointing locations from space in the same decade.

The sheer volume of data now being stored and processed digitally is incredibly vast. In 2010, there was astonishment when the International Data Corporation revealed that two zettabytes of data were being created and replicated worldwide each year. (A zettabyte is equivalent to 10^{21} bytes—that is,

1,000,000,000,000,000,000,000,000 bytes. Physically storing one zettabyte of data would require about 250 billion DVDs.) By 2025, annual global data creation is projected to surpass 180 zettabytes.[37]

Improvements in data collection, storage, and analysis have helped power many of the greatest technological advances of this century. Data is now playing a critical role in the development of machine learning, predictive analytics, and AI, in addition to technology such as facial recognition and natural language processing, which can augment business intelligence by making it easier for enterprises to "listen in" on cultural conversations. At the same time, the explosion in data has also brought about the urgent need to protect it from theft or exploitation, through cybersecurity and privacy legislation.

Data science used to be a niche department in a handful of universities, but the discipline is now valued worldwide by businesses. The arcane principles of data science are being democratized with AutoML (automated machine learning) to allow laypeople to access its benefits. Likewise, "small data" applications are being developed to "facilitate fast, cognitive analysis of vital data in situations where time, bandwidth or energy expenditure are of the essence."[38] Think self-driving cars that may need to make quick, life-saving decisions when an emergency prevents them from relying on access to a cloud server.

Data is also fueling the evolution of AI, IoT, cloud computing, and superfast networks like 5G, which are themselves energizing the development of smart homes, factories, and cities.

How can we make more out of advances in data? According to McKinsey's vision of the data-driven enterprise of 2025, proactive businesses will "find ways to use data to optimize nearly every aspect of their work." We can expect challenges that previously demanded multiyear roadmaps to be resolved in weeks, days, or even hours. At more and more businesses, chief data officers will become among the most important members of the C-suite, leading critical business units with profit-and-loss responsibilities.[39]

As you can see, in today's world, harnessing the crucial elements in the Formula for Good is more complicated—and more urgent—than ever.

Find Your Own Formula for the Future

Companies that don't harness the Formula for Good can lose their bearings and sense of perspective. You can be sucked in by false promises of an easy path to sustainability. You think you're on the right street, but actually you're walking into the dangerous part of town. It's unintentional, but without the right expertise, you're listening to the wrong opinions and failing to identify current opportunities. Without a proactive technology strategy, you will stand still while competitors advance. And without a holistic view of data, you can end up focusing on one or two strengths but miss the weaknesses that are letting the business down.

When businesses fail to grapple honestly with the challenges of sustainability, confirmation bias creeps in, whereby you interpret new evidence as reinforcement of what you already think. The opportunity is lost. The shortcomings are ignored. It's like the parent who proudly announces that little Johnny got an A in physical education but forgets to mention that he flunked math and English. That blinkered view may help the family feel good about Johnny in the short run, but in the long run it will likely spell disaster.

By contrast, those who embrace conflict, acknowledge their weaknesses, and challenge the status quo regularly may find themselves out of their comfort zone, but the rewards are shown in greater awareness of current trends and more responsive innovation.

I don't know what the future will bring, but I'm confident that we'll need a combination of expertise, technology, and data to figure it out and respond to the challenges. Whatever shape the Fifth, Sixth or Twentieth Industrial Revolutions take, we have a formula that will help us navigate them.

In the next chapter, we'll take a deep dive into the world of mobility to see how the Cycle of Good and the Formula for Good are playing out across some of the world's biggest and most important industries, from automobiles and aviation to space exploration and smart cities.

Key Takeaways

- The Covid-19 vaccination development and the growing awareness of the need for environmental innovation offer powerful examples of the Formula for Good.
- From the steam engine to the digital bit, innovation has been driven by a combination of expertise, technology, and data—all guided by a Noble Purpose.
- This Formula for Good continues to power many innovations in our modern lives, creating benefits for millions worldwide.
- Businesses today need to harness advances in tech and data—although there is a talent shortage globally.
- Whatever the future holds, the Formula for Good offers a road map for sustainable success.

THE INVENTION TEST: PART 2

N ow it's time for the second installment of our quiz on some of the most notable sustainability innovators of all time. How many of the six questions that follow can you answer correctly? Turn the page and find out.

Question 7

Whose most famous appliance was inspired by living and working with the hearing impaired?

A. Alexander Graham Bell
B. John Logie Baird

The answer is A: Alexander Graham Bell

Scottish-born Alexander Graham Bell is well known for his invention of the first practical telephone, which had a seismic impact on society in the last quarter of the 19th century—and beyond.

Bell's interest in sound technology stemmed from his personal experience of communicating with his mother and wife, both of whom were hearing impaired. His father, a speech elocution professor, developed a kind of "visible speech" that used symbols to help deaf children talk. The younger Bell became a professor himself, improving on his father's system when the family moved to Boston, where he married one of his university students. The *decibel*, a measure of the pressure exerted by a sound wave, is named after Bell.

Whether Bell was the *first* to make a working telephone is open to debate—another inventor, Elisha Gray, developed a similar design at almost the same time. However, Bell was certainly the first to secure a patent for the invention, which he defended for the rest of his life. Amusingly, Bell refused to have a phone in his study, as he found it annoying. But not annoying enough to stop him from founding the Bell Telephone Company, which later became AT&T—still among the world's largest telecommunications companies.

However, the inventor didn't get it all his own way. He was determined that people should answer the phone with the greeting *ahoy*, while Thomas Edison (a fast follower where telephones are concerned) favored *hello*. Edison prevailed. This historical tidbit explains why the old-fashioned Montgomery Burns uses the expression *ahoy, hoy* in *The Simpsons*.

John Logie Baird was another Scottish inventor and the first to demonstrate television in 1926.

Question 8

Which company topped the list of organizations with the most U.S. patents granted for 29 consecutive years from 1993 to 2021?

A. IBM

B. 3M

The answer is A: IBM

For over 130 years, IBM has been at the forefront of computing advancements. From its early days in punch-card data processing to the development of magnetic tape, hard drives, floppy disks, and personal computers, the company has been a byword for innovation. In 2022, its 8,682 patents included plans for second-generation, two-nanometer nanosheet technology that can house 50 billion transistors in an area about the size of a fingernail. (Two nanometers is smaller than the width of a single strand of human DNA.)

The 29 consecutive years of IBM's patent leadership began with IBM's appointment of Lou Gerstner as CEO, who shepherded the company back to growth after a period of decline in the 1980s.

If you went for 3M, you weren't far off. The company's track record in innovation over the last century is also impressive, including a portfolio of 60,000 products and more than 130,000 patents worldwide. 3M's "15% Culture," which permits employees to dedicate 15 percent of their work time to trying out cool new stuff, led to the Post-it note, among many other winning inventions.

Samsung out-patented IBM in 2022, then retained the top spot in 2023.

Question 9

When was the first surgical procedure conducted under anesthesia?

A. 1684
B. 1846

The answer is B: 1846

In 1684—and for millennia before that—all surgical operations, from pulling teeth to limb amputations, would have been excruciatingly painful. Efforts were made to numb the patient's pain through intoxication (hard liquor), sedation (opium), or even knocking them unconscious. But generally, the procedure was agonizing, forcing the surgeon to work as quickly as possible in an effort to minimize the suffering.

The development of experimental science during the Enlightenment of the 18th century led to research into gases such as nitrous oxide (laughing gas) and then ether—both of which became riotously popular as recreational drugs. Users also noticed their pain-alleviating qualities.

In the 1840s, two Boston dentists began testing the use of nitrous oxide and ether as painkillers on themselves and then their patients. While Horace Wells investigated the use of nitrous oxide, William Morton succeeded with ether.

In 1846, Morton participated in the first public demonstration of ether as a general anesthetic during an operation to remove a neck tumor at Massachusetts General Hospital. Ether remained the anesthetic of choice in the U.S. for the rest of the century. In the U.K., however, chloroform was preferred, partly because Queen Victoria used it during the birth of her eighth son, Prince Leopold, in 1853.

Question 10

Which company founder said, "The day has passed when business was a private matter—if it ever really was. In a business society, every act of business has social consequences"?

A. Robert Wood Johnson (Johnson & Johnson)
B. Samuel Curtis Johnson (SC Johnson)

The answer is A: Robert Wood Johnson

During the American Civil War, about two-thirds of the 720,000 reported casualties were caused by infection and disease. Contagious illnesses such as typhoid and malaria were rampant, and amputations after battlefield injuries were hurried and dirty, leading to dangerous infections. The young Robert Wood Johnson was deeply affected by stories from the front lines. His family business was already making medicated bandages when he became inspired by Dr. Joseph Lister's radical new procedure: antiseptic surgery. (Lister was himself inspired by Louis Pasteur's germ theory.)

Johnson went all in. In partnership with his two younger brothers, the manufacturer innovated the world's first mass production of sterile surgical supplies to reduce the rate of infection. Johnson & Johnson would grow into a multinational pharmaceutical, biotechnology, and medical technology corporation that today is one of only two U.S. companies to hold a prime AAA credit rating (the other is Microsoft).

The household products company SC Johnson—unrelated to Johnson & Johnson—showed its own values-led approach in 1975, when founder Sam Johnson vowed to eliminate CFCs from its aerosol products worldwide, taking a public stand against chemicals that were later banned for damaging the environment.

Question 11

How old was Louis Braille when he started developing his tactile alphabet for the visually impaired?

A. 15 years old
B. 91 years old

The answer is A: 15 years old

Louis Braille was blinded at the age of three while playing with an awl in his father's harness shop. He became an accomplished organist and began attending the Royal Institute for Blind Youth in Paris in 1819. The only way he could read was by tracing whole letters and numbers embossed on paper (the so-called Haüy system, named after the founder of the institute). Books produced in this way were expensive and clumsy to use, leading Braille and others to search for an improved system.

At age 15, Braille took an interest in a 12-dot cryptography system created for soldiers to communicate in the dark on the battlefield. Louis adapted it into a six-dot code that could be read using a single index finger. Poignantly, to make the dots, he used an awl that was similar to the one that had blinded him as a boy.

Braille's code took several decades to become popular, partly because some teachers feared that they would lose their jobs if the blind could read and learn for themselves. Today, the system is used globally in all languages and is incorporated into many assistive technologies, but always employing the same six-dot design that Braille pioneered.

Question 12

Which rock band known for their innovative use of music technology was previously named the Screaming Abdabs?

A. Pink Floyd
B. Led Zeppelin

The answer is A: Pink Floyd

T he band also performed as Sigma 6, The T-Set, and The Megadeaths, until front man Syd Barrett suggested combining the first names of his favorite blues singers, Pink Anderson and Floyd Council.

Pioneers of progressive rock, Pink Floyd broke new ground in music technology with their revolutionary use of sampling, stadium lasers, and the invention of the Azimuth Coordinator, which surrounded the listener in quadraphonic live sound. Unable to read sheet music, the band also devised their own system of notation.

Pink Floyd is only one of the latest in a long line of innovators who applied technological breakthroughs to stimulate creativity in the art of music. Italian Bartolomeo Cristofori designed the first pianoforte in 1700. Thomas Edison's 1877 phonograph brought recorded music into homes, while Les Paul's 1941 electric guitar changed sound forever, making possible genres including modern jazz, blues, and rock. With the application of synthesizers, turntables, personal computers, and now AI to music, the interaction of art and technology shows no signs of slowing down.

4

MOBILITY FOR GOOD

- The rise, decline, and future of the railroads
- How the car and plane changed the world . . .
- . . . and must now rise to their greatest challenge
- What's the fuel mix of tomorrow's mobility?
- Why smart cities can benefit society and the planet

"To do more for the world than the world does for you—that is success."
—Henry Ford

Transportation typifies the endeavors of business to harness innovation for the progress of society. Mobility leaders are some of the biggest companies in the world. As widescale employers and important contributors to local and international communities, they face urgent challenges that will impact their continued success. And as major producers of carbon emissions, the choices they make today will have a long-lasting impact on life on Earth in the decades ahead.

In this chapter, we'll look at companies engaged in rail, automotive, aviation, shipping, and the rise of smart cities as Forces for Good in the world—and how pioneers have used a combination of expertise, technology, and data to meet the cultural needs of their era.

Miracle and Fall

The transformation of America in the 19th century from a patchwork of rural communities into a global, industrial powerhouse is inseparable from the story of the railroads.

"America was made by the railroads," wrote historian Christian Wolmar in *The Great Railroad Revolution.*[1] "They united the country and then stimulated the economic development that enabled the country to become the world's richest nation. . . . Quite simply, without the railroads, the United States would not have become the United States. The extraordinary growth of the railroads changed the very nature of America."

The railroads represent a feat of innovation astonishing enough to rival any in history, setting almost two hundred thousand miles of track across all terrains, over the course of 70 years.

The Industrial Revolutions in Europe, especially in Britain, had greatly advanced the technology of locomotives, carriages, and the rails to carry them. But the vast, open spaces of the U.S. posed a special challenge for ambitious entrepreneurs as they sought to apply the technology in this huge, developing country. Waterways and toll roads, once the supreme modes of transportation in American, gave way to rail as the transport of choice for both freight and passengers. The iron horses on their iron roads soon rode supreme.

The growth of the railroads was sparked by the frontier mindset of restlessness and determination, and they soon began influencing culture in turn, connecting towns along the Eastern Seaboard and powering their growth into thriving economic centers. By the 1840s, the fast followers were piling in, bringing speculative inventions and investments to accelerate the rate of progress.

In the early years of expansion, the federal government largely kept its distance, while occasionally supporting railroad companies with tax exemptions and land grants. Landmark projects such as the nearly 450-mile New York and Erie Railroad—briefly the longest in the world—showed the potential for opening up trade routes by carrying goods from the Hudson River to the Great Lakes. The construction team faced a wide range of challenges, including labor disputes, financial setbacks, severe terrain, and stakeholder opposition, especially from the operators of the Erie Canal, who stood to lose out. Little wonder that the opening ceremony in 1851, led by President Millard Fillmore, went down in history for its extravagance. It was truly cause for celebration.

Achievements like the Pennsylvania Railroad (which grew to become the largest corporation in the world and second in the U.S. only to the federal government itself in terms of budget) set the tempo for the remarkable 1,776-mile transcontinental railroad in 1869, originally greenlit by President Abraham Lincoln to symbolize the nation's coming together after the Civil War. The telegram from Promontory Summit, Utah, to New York confirming its completion at 3:05 p.m. on May 10, 1869, read as follows: "The last rail is laid; the last spike driven; the Pacific Railroad is completed. The point of junction is 1,086 miles west of the Missouri River and 690 miles east of Sacramento City."[2]

The transcontinental railroad proved much more than a symbol of national cohesion. Journey times from East to West, previously undertaken by horse and wagon, or the long way around by boat, were reduced from over a month to less than a week.[3] California advanced from backwater to mainstream. The cost of travel plummeted, allowing less wealthy Americans to reach parts of the country they had thought beyond their means. Time itself became systematized, as the railroad managers introduced the system of four time zones (Eastern, Central, Mountain, and Pacific) to avoid the scheduling confusion caused by a multitude of local times.[4]

By the 1880s, the railroads were the largest national employer outside of agriculture, amassing a fleet of around 40,000 locomotives, split evenly across freight and passengers.[5] Pullman cars transported the masses in relative comfort. Rail gauges were standardized to allow rival companies to link up into a national network. Mechanized farming was improving crop yields, and the railroads carried this bounty to the fast-growing cities, lowering the cost of food and other staples. American engineering was now the envy of the world.

The way of life before the railroads was now a mere memory. Alongside the economic changes, most obviously in the form of jobs and new towns, the iron roads had made possible new concepts like mail-order shopping and same-week postal deliveries. Train stations rose like cathedrals in glass and shining steel. In major cities, skyscrapers full of white-collar workers reached upward around the main terminals.

Telegraph lines ran parallel with the tracks in a symbiotic relationship, taking advantage of the railroads' right of way while providing rapid communications to aid scheduling. In 1866, a reliable transatlantic cable connected Britain to the U.S., allowing messages to pass from London to New York in just a couple of minutes. The world was getting smaller again.

Perhaps the greatest gain was intangible: confidence. There was an electric current of possibility that crackled along the railroads. And this surging optimism ultimately invited novel innovations in mobility that would surpass the railroads in the following century.

American Railroads—Back on Track?

Today, the U.S. railroad network remains the world's longest network and a critical transporter of national freight. Yet it lags behind other nations in Europe and Asia for passenger transportation, with a freight-to-passenger ratio of 5:1 (Europe is the inverse).[6] This is partly down to geography and population density, but the U.S. passenger network has also suffered in terms of both investment and prestige over the last century.

In Japan, the Tokaido Shinkansen bullet train covers the 247 miles from Osaka to Tokyo in just two and a half hours at speeds of 177 mph, arriving with split-second precision. Blink and you'll miss Mount Fuji. The journey time is expected to drop to nearly one hour when the 310-mph L0 Series is introduced in the next 20 years.[7] Services are clean, frequent, and comfortable. Using maglev (magnetic levitation) technology, the L0 starts to "float" at speeds exceeding 93 mph. Maglev trains in China reach similarly high velocities, as do electric trains in France, Italy, and South Korea. But American railroads offer no such speed and convenience—although plans to build a superconducting maglev train through the Northeast Corridor of the U.S., connecting Washington, D.C., to New York in less than an hour, are currently under review.[8]

In the competition for market share with road and air, U.S. passenger rail companies need to draw on the Formula for Good: expertise, technology, and data. They might draw inspiration from the 4C Challenge, as laid out by the Railway Technical Strategy, a planning document developed by a consortium of U.K. industry leaders: "How to improve *customer* experience whilst increasing network *capacity* and at the same time reducing both *carbon* footprint and the *cost* of running the railway?"[9] Achieving all four goals described by the four Cs is a delicate balancing act. By focusing too much on one, operators will inevitably detract from the rest.

Digital innovation is the golden thread across all four challenges. For example, greater internet connectivity, such as through 5G wireless, improves the customer experience, helping passengers feel comfortable in an extension of their home or office. At the same time, internet-based tools can help railroads increase their passenger capacity. To attract traffic away from the roads, operators need to increase the frequency of trains. Digitalization and automation of trains and rail signaling to ensure optimal emergency stopping distances can bring more services per hour onto the tracks in total safety. Onboard sensors and Internet of Things devices can help to anticipate failures and diagnose trains' health to avoid delays through preventative maintenance and the predictive delivery of spare parts. These benefits can reduce costs, since around 40 percent of the operating costs of a train are linked to its equipment maintenance over a 35-year lifetime. What's more, so-called digital twins—simulations of entire train systems using real-time data to model their behavior—also allow operators to monitor trains without interrupting services.[10]

The environmental and social benefits of a popular railway network are evident in numerous countries around the world. Indeed, 75 percent of Americans agree with the statement that "more trips should be shifted to passenger rail and public transit to address the impacts of transportation on climate change."[11]

Rail has an enviable position among its rivals as a more environmentally sustainable mode of transport. The shift to electric trains powered by renewable energies can help reduce the consumption of diesel fuel, while new fuels such as hydrogen and liquefied natural gas are currently being tested in France and Spain. Can rail work with other industries to fast-track the development of these new fuels? The challenges to rail are significant—but the Formula for Good suggests a pathway to a renaissance of train travel in the U.S.

Pedal to the Metal

The birth of both the automobile and the airplane at the start of the 20th century reflected the cultural mood of the time. As the so-called American Century dawned, U.S. industry was booming, built on the railroads, steel, oil, and the modernization of agriculture. Citizens were feeling increasingly prosperous and open to new means of raising their standard of living still further. The middle classes were on the move.

The country still faced social challenges, but there was at least a growing sense of peace, and the trauma of the Civil War was receding. The Spanish-American War of 1898 had proved successful, creating national pride in the past and confidence in the future. A hero of that conflict, the newly elected president Teddy Roosevelt, embodied the nation's youthful exuberance and can-do spirit. At just 42 years old, he was the youngest person ever to occupy the office, a distinction he holds to this day.

A spirit of optimism permeated trade and politics, entertainment and culture. World's Fairs were opening eyes to other realms, stimulating a desire to explore and seek adventure. Jack London stirred the imagination with his bestselling novel *The Call of the Wild*, while silent films and the mass print media gave Americans a different perspective on their everyday lives. Sodas like Dr. Pepper, Coca-Cola, and Pepsi-Cola were slugging it out for market share. Inventors like Thomas Edison and Nikola Tesla were household names.

In the fledgling automotive industry, fresh innovations from Germany were ripe for development. In 1876, Nicolaus Otto had invented the gasoline-powered internal combustion engine. By the 1880s, engineers Gottlieb Daimler and Karl Benz had harnessed its power to pull stagecoaches. Their rival workshops would eventually combine to form Mercedes-Benz. Throughout the rest of the world, car manufacturing became a craze, with hundreds of small shops building a handful of machines a year. Many brands fizzled out, but others like Peugeot, Renault, Fiat, and Rolls-Royce put down deeper roots. Bicycle makers like Rover, Morris, Skoda, and Opel saw the opportunity to pivot from two wheels to four.

In the U.S., established manufacturers with the necessary facilities, tools, and manpower could hit the ground rolling. Studebaker and Durant had previously built horse-drawn vehicles; Olds made stationary gas engines; Buick cast bathtubs. All were able to convert some of their manufacturing equipment to the new automotive industry.

Henry Ford had no such head start. He was a decent engineer and held a prominent role at the Edison Illuminating Company in Detroit. In his spare time, the young entrepreneur built a motorized quadricycle in his coal shed. But it created few waves. Even with the backing of Edison and several investors, his first attempt to form the Ford Motor Company failed after the workshop struggled to meet demand. He could have easily ended up as one of those many dreamers who were forgotten to history.

But Ford was no ordinary dreamer—and he had no ordinary dream. "I will build a car for the great multitude," he said. "It will be large enough for the family, but small enough for the individual to run and care for. It will be constructed of the best materials, by the best men to be hired, after the simplest designs that modern engineering can devise. But it will be so low in price that no man making a good salary will be unable to own one."[12] From the start, he saw the greater good and had a Noble Purpose to match.

Ford set up again and launched his iconic Model T in 1908. Ford sold more than 15 million Model Ts over the next two decades, both at home and abroad. By 1918, half of all cars in the U.S. were "Tin Lizzies," available, as Ford joked, in any color the customer wanted—so long as it was black.[13]

The game-changing success of the Model T held true to the Formula for Good, combining technology, expertise, and data. Ford wanted to offer a ride that was economical but exceptionally rugged so it could withstand the poor state of unsurfaced roads at the time. He therefore pioneered the use of vanadium steel, which was both light and tough. For the tires, he tested new forms of rubber with Harvey Firestone and Thomas Edison.

Ford was happy to bask in the limelight of success, but he later echoed Andrew Carnegie, saying, "I don't do so much, I just go around lighting fires under other people." But he happened to have some very capable thinkers and doers to rely on.

In terms of expertise, William Knudsen and Charles Sorensen, both originally from Denmark, kept the wheels turning ever faster during Ford's years of expansion. With renowned "problem solver" Clarence W. Avery, they oversaw the development of the modern assembly line and the first true mass production in 1913, which revolutionized manufacturing far beyond the automotive industry. In the earliest days, Sorensen claimed that he harnessed himself to a chassis like an ox and pulled it past workers to demonstrate the assembly line concept. The line was broken down into 84 steps, and workers were trained to do just one step at maximum speed and quality. A mechanized conveyor belt—not an ox—pulled the chassis through the factory, accelerating to a top speed of six feet per minute. Parts were stamped out automatically by custom-built machines. European manufacturers were amazed to see that Model T parts were standardized and therefore easily interchangeable.

Avery's innovations in time and motion management, captured with stopwatches and extensive data analysis, reduced the times for building a car from 12 hours to one hour and 33 minutes.[14] The reduction didn't happen all at once but as the result of continual, painstaking improvement accompanied by detailed analysis and further change. As Avery explained, "A well-understood failure is better than a misunderstood success."[15]

Thanks for Ford's innovations, the price of the Model T dropped from $950 in 1909 to $360 in 1916 and eventually down to $290 in 1926.[16] "There is one rule for the industrialist and that is: make the best quality of goods possible at the lowest cost possible, paying the highest wages possible," wrote Ford, and he was good to his word.

At the same time, Ford was highly aware that his enterprise needed to offer real value to his employees, who grumbled that the assembly line made their work monotonous. He raised their pay to $5 per day, at a time when the average wage for similar workers was $2, to keep them motivated—which also gave them the spending power to buy a Model T of their own. Unsurprisingly, Ford faced criticism from other business owners. But he earned a reputation as a visionary who looked beyond his own industry to the impact he could have on society at large. By turning the car from a plaything of the rich to a utilitarian vehicle for the growing middle classes, Ford left an indelible mark on social history.

"A business that makes nothing but money is a poor business," he later wrote. "To do more for the world than the world does for you—that is success."

The business innovations pioneered by Ford made a powerful and lasting contribution to American industry. In the words of Mark Fields, former CEO of Ford and of Hertz, "Henry Ford believed that a good business makes excellent products and earns a healthy return. But he proved that a great business does all that while creating a better world."

Administrative Genius

The early decades of the 1900s were a period of intense competition to define the automobile industry, creating a continual and changing stream of winners and losers along the way. In time, even Ford, the great innovator, fell behind the pace of change. He became too entrenched in his mantra that *less is best* and clung stubbornly to the original blueprint of the Model T, recognizing too late that the wheel of culture had spun. The middle classes that he helped to mobilize were now hungry for vehicles with a bit more style than boxy black.[17] In the Roaring Twenties, Americans aspired to curves and glamour—and some growl beneath the hood.

Ford's competitors had listened to the music. Under the leadership of William C. Durant, General Motors (GM) had grown in Ford's shadow, consolidating several companies like Buick, Oldsmobile, Cadillac, and Oakland (later Pontiac), which built more upmarket, sleeker models. At first, GM couldn't compete with the popularity and price of the Model T . . . but culture doesn't stand still.

By the 1920s, Durant had been pushed out of the leadership of GM, and an engineer named Alfred P. Sloan assumed control, appointed president in 1923 and holding the post for the next two decades. William Knudsen had also jumped ship from Ford to GM, bringing his vast expertise in mass production. GM tapped into his knowledge to become a fast follower of Ford's technological advances in assembly-line manufacturing.

Widely described as an administrative genius, Sloan tightened GM's loose managerial structure and greatly strengthened the company's ragged sales organization. Proud to be called a white-collar man, Sloan decentralized production but centralized administration, setting a managerial style that is still copied today.[18]

While Ford played it safe by manufacturing Model Ts virtually unchanged for almost two decades, Sloan masterminded annual style changes in car models and a pricing structure that tiered Chevrolet, Pontiac, Oldsmobile, Buick, and Cadillac from least to most expensive. Not only did this reduce friendly fire between allied brands, but customers could upgrade from one GM nameplate to the next as their budget and needs changed. "A car for every purse and purpose" was a ladder of desire that both encouraged and satisfied aspiration.[19]

Sloan also pioneered advances in customer credit that would establish the foundations of a whole new financial industry.[20] The risk-averse Ford would only accept cash. With Sloan's loans, customers who had saved for a Model T could now purchase something jazzier—and in different colors, too. As customers traded in their old cars for the next year's must-have, Sloan the administrator was ready to maximize returns on the glut of used cars.

Sloan didn't stop with U.S. dominance. He masterminded GM's foreign expansion, adding overseas operations through purchases of companies like Vauxhall (U.K.) in 1925, Adam Opel (Germany) in 1929, and Holden (Australia) in 1931. GM also grew its refrigeration sideline, Frigidaire, into a market leader under his watch. Where Ford lit fires beneath his people, "Silent Sloan" liked to work in the background, devolving power and rewarding collective performance. "The ability to get people to work together is of the greatest importance," he wrote in his 1963 management guide *My Years with General Motors*, which remains a bible for business students worldwide.

A doyen of data, Sloan professed a "factual approach" to making decisions:

> The final act of business judgment is of course intuitive. . . . But the big work behind business judgment is in finding and acknowledging the facts and circumstances concerning technology, the market and the like in their continuously changing forms. The rapidity of modern technological change makes the search for facts a permanently necessary feature of the industry.[21]

In other words, put your trust in technology and listen to the data.

By 1927, Ford belatedly accepted that the era of Model T dominance was over and released the Model A. Of course, that wasn't the end of Ford. Far from it. The brand prospered in the decades ahead. But GM was now the industry leader. Sloan's management of technology, data, and expertise had won the decade.

By 1930, these big two had become the "Big Three." Walter P. Chrysler was a railroad man before switching to cars at Buick. Having built up the brand as part of the GM family, he went solo, unveiling the Chrysler Six in 1925 to considerable acclaim. His big move came in 1928 when the young business bought out the established car and truck maker Dodge Brothers. As the Chrysler Building rose in New York to become (briefly) the world's tallest building, the company launched the Plymouth and DeSoto brands to meet demand in the low- and mid-market categories, respectively. The Great Depression weeded out many of the remaining smaller brands, leaving the same three giants to tough it out at the top of the U.S. auto market for the next 80 years.

Into the Fast Lane

While the rise of the automobile was itself generated by culture, few innovations have had such a marked impact on living habits and social customs. The scale of change stimulated by the rise of the automobile is incalculable.

Cars mobilized suburbanization, leading to new towns, a construction boom, and higher living standards. As the largest companies in the world, car manufacturers pioneered new concepts in mass marketing and consumer credit. In industry, too, the rise of this brand-new sector created a massive demand for steel, rubber, petroleum, leather, paint, glass, and other materials as the decades passed, creating whole sub-industries to supply the manufacturers. Secondary industries such as repairs and parts took hold, as did motor services like dealers, gas stations, and motels.

Even the diet of the nation changed, fueled by diners with their fast-food hamburgers, French fries, milkshakes, and donuts to go. Travel, tourism, and vacations were now within reach of millions of middle-class and working-class Americans. There was freedom from the grime and smoke for urban families and bright lights and department stores for out-of-towners. Dance halls and cinemas swelled in number, making stars of musicians and actors. Young people gained a taste of freedom, away from the parental gaze.

Government intervention facilitated the expansion of the automobile in the form of new roads, with subsidies provided by the 1916 Federal Road Act. President Woodrow Wilson—himself an enthusiastic motorist—endorsed the law in the Democratic Party platform of 1916, which observed, "The happiness, comfort and prosperity of rural life, and the development of the city, are alike conserved by the construction of public highways."

Of course, the rise of the automobile wasn't all good. Cities now experienced traffic jams and exhaust pollution. But there was no putting the cork back in the bottle. In fact, things were about to get a whole lot bigger.

The economic boom following the Second World War marked the heyday of the automobile in the 1950s. Car ownership shot up, with at least one model on virtually every suburban driveway. Highways improved, and the service industries mushroomed.

The automobile industry mirrored culture. Collars, pleats, and car engines all expanded, and fins, convertible tops, and chrome plating captured the exuberance of 1950s style. Elvis bought a pink Cadillac and sang about it. James Dean's Mercury Coupe helped make teenage angst cool in *Rebel Without a Cause*. Sports cars, led by the Chevrolet Corvette, brought the racing track onto Main Street. Supercars from European makers like Ferrari, Mercedes, and Rolls-Royce were now cruising above 100 mph and creating a new market segment for luxury and opulence.

An Act of Altruism—and Smart Business

Of course, the cycle of culture continued to turn. By the 1960s, family cars were becoming more modest and utilitarian. There were still style icons like the Mini Cooper, the Ford Mustang, and the smooth Aston Martin DB5 (the chosen car of iconic spy James Bond), but most car designs were becoming increasingly economical.

The 1960s were also a period of increased safety in cars, with regulations shifting responsibility from the driver to the maker. Again, business that rose to the challenge. GM began development of its Hybrid series of crash test dummies, which would become the industry standard for use in improving auto designs. The manufacturer shared its technology with government regulators and the car industry. Today, the Hybrid III is the "most popular anthropomorphic test device in the world for the evaluation of automotive safety restraint systems in frontal crash testing."[22]

The Swedish manufacturer Volvo was ahead of its time in recognizing the growing demand for safety as an opportunity to stand out in the market. And, yet again, this was a story of expertise, technology, and data. Volvo's president Gunnar Engellau had lost a relative to a car accident, partly due to the flawed design of then-widely-used two-point seat belts (similar to an airplane lap strap). In 1955, the high-profile death of James Dean, who might have survived had he been wearing a seat belt in his Porsche 550 Spyder, also raised safety awareness among the public. Determined to stop others from experiencing this tragedy, Engellau poached an engineer named Nils Bohlin from Swedish rival Saab and set him to work on finding an alternative.[23]

Bohlin understood that two-point belts were not only flawed from a safety perspective, but they were also awkward, uncomfortable, and un-macho—and therefore rarely worn. His revolutionary three-point seat belt was not only more comfortable, but also easier to fasten one-handed. The straps also helped to protect the pelvis and upper torso.

Volvo ran hundreds of experiments and researched tens of thousands of accidents to test its invention. The results confirmed that Bohlin's three-pointer would save lives and lessen injuries during accidents—which it has done for the last 50 years for millions of drivers.

Volvo became the first auto maker to offer seat belts as standard equipment. But Engellau believed the three-point seat belt was too important for society to limit its availability. Therefore, rather than register a patent on the new design, Volvo shared the technology freely with the wider automotive industry.

This act of altruism also bore the hallmarks of realism. The culture wasn't yet ready for his innovation. Seat belts were unfamiliar and therefore actively disliked by many drivers despite the evidence that seat belts saved lives. Just one in ten drivers in the U.S. chose to belt up until legislation and widespread publicity caused a cultural shift in the 1980s and 1990s.

Now, across the world, billions of people use Bohlin's invention every day. It's worth adding that even now, eight percent of Americans still refuse to wear a seat belt.[24]

Volvo's business did benefit greatly from its status as a builder of safe cars. Bohlin would later help design the first side impact protection system, which burnished his company's reputation still brighter. More recently, safety standardization has raised levels across the board, eating into Volvo's unique reputation for safety. Not a brand to stand still, Volvo has widened its safety orientation to include the environment in addition to its drivers, fulfilling its pledge to ensure that every new Volvo car launched from 2019 onward would have an electric motor—either fully electric or hybrid.[25] The brand aims for 50 percent of car sales volume to be fully electric by 2025.

More Than Car Companies

Over the past 40 years, the U.S. auto industry has faced several setbacks, including, at times, a slowness to innovate in step with society's needs. The arrival of international brands in the U.S. market and the growth of longer-lasting models have contributed to the decline in American auto manufacturing from around 10 million cars per month in the 1970s to 1.4 million in 2021.[26] The Big Three's market share in the U.S. has fallen from around 85 percent in the 1960s to just under 40 percent today.[27]

Yet the automotive industry remains an integral part of the U.S. economy. Nearly a million Americans work in motor vehicles and parts manufacturing, and over 1.25 million are employed by dealerships. Over three percent of America's GDP stems from the industry. While the Covid-19 pandemic provided a major bump in the road, current trends point to a speedy recovery.[28]

From a social perspective, the stories of the Big Three companies are uniquely tied to their communities, with several generations of the same families working in their vast factories. Whole towns exist because of the auto industry, creating a sense of responsibility towards those constituents.

Newcomers have looked to shake up the industry over the years. In the 1970s, John DeLorean had a vision for an "ethical" car that was sporty, safe, and sustainable.[29] A respected engineer at GM, he attracted plenty of financial backing. But the project was beset with technical problems from the start and would end as one of the most storied flops in automotive, if not broader industrial, history.

DeLorean's timing was all wrong. Society wasn't ready to buy into his vision of the future. The design of his car—familiar to movie fans from its appearance in the *Back to the Future* trilogy—was futuristic and appealing, but the technology was flawed. As a result, the value given to the enterprise from stakeholders such as investors, employees, and the British government (which provided a subsidized base in Northern Ireland) greatly outweighed the value they received from DeLorean.

The electric vehicle (EV) company Tesla Motors has also aimed to bring an ethical alternative to the market in the 21st century. Where DeLorean failed, Tesla has grown into the most valuable auto brand in the world, with a market capitalization that is close to twice the size of its nearest rival, Toyota.[30]

The history of brief highs and costly lows that have haunted EV development over almost two centuries could have persuaded the Tesla founders to invest elsewhere. The steady stream of false dawns in electric auto manufacturing dates back to the early 19th century, with any number of high-profile casualties strewn on the side of the road. Even as late as the 1990s, impressive innovations such as GM's EV1 failed to gain market traction. Environmentalism was still too niche, and eco-warriors didn't have the cash for high-performance cars. American culture simply wasn't ready to embrace electric.

From the start, the Tesla strategy was different. The design was classic retro rather than futuristic. Despite the inevitable glitches that came from building a supercomputer on wheels, the powerful lithium-ion batteries provided a drive that compared favorably with gas cars at the same price point. Even the marque, honoring the inventor Nikola Tesla, who first patented the alternating current (AC) induction motor in 1888, was deliberately non-eco and non-tech sounding.

The business model was also different from previous electric contenders, which aimed to saturate the market with affordable, mass-produced highway buggies. Tesla chose to create a truly desirable and stylish car that would appeal to both conscientious drivers and those who wanted to stand out from the crowd. Once the brand was established and the image of EVs was changed, Tesla would then move into more accessible consumer markets.

Making Your Own Luck: Timing Is Key

The Tesla proved to be the right car at the right time. The mood had evolved in the 15 years since the demise of the EV1. On the boulevards of Hollywood, the hybrid Toyota Prius had enhanced the prestige of driving an environmentally friendly car at a time when Americans were increasingly nervous about the price of gas. Climate change was entering everyday conversation, and the importance of sustainability was increasingly discussed in boardrooms. There was a growing appetite for living green, even if most people still didn't want to admit it. If only companies could make decarbonization cool . . .

When it went on sale in 2008, Tesla's Roadster made a big splash. "This is not your father's electric car," cheered the *Washington Post*.[31] "The $100,000 vehicle, with its sports car looks, is more Ferrari than Prius—and more about testosterone than granola. . . . Bottom line: Electric cars don't have to be for wimps."

"It is not just a car, but one of the strongest automotive statements on the road," concluded *Car and Driver*.[32]

The Model S delivered further on the hype, outselling its luxury sedan rivals and scooping up major awards that were traditionally the preserve of gas cars.[33] The vehicle extended the range, with higher acceleration speeds and over-the-air software updates that improved performance automatically.

The arrival of the Model 3 in 2017, with its starting price tag of around $40,000, achieved the vision of providing a more accessible car that was powerful, beautiful, and produced zero tailpipe emissions.

Constant Disruption

T hose early EV models had their critics, but they brought about the trans- formation they were designed to initiate. Electric cars no longer looked like sci-fi bubbles or golf carts. Tesla showed that being eco-aware could be sexy, providing a boost to industries beyond the auto world. It reinforced a growing sympathy for environmentalism and helped to establish a new nor- mal of cleaner energy. As a first mover, it seized on a cultural vibe and turned that feeling into commercial demand.

Working from a blank sheet of paper, Tesla could be a different kind of car company. From the start, it saw the importance of owning the battery and charging space as much as the EV itself. In 2014, Tesla "did a Volvo" by open- sourcing its battery patents, with the aim of fast-tracking mass adoption.[34] The gesture has helped scale the battery energy storage industry globally. In 2010, one kWh of battery capacity cost $1,200. The cost is expected to drop to $80 by the end of 2026.[35]

Led by Tesla, supercharger stations are now becoming widespread across the U.S., with many provided in safe, well-lit, amenable locations—a compel- ling factor in buying decisions. Buoyed by government incentives, rival man- ufacturers are increasingly cooperating to create a nationwide, fast-charging plug-in network available to all EV drivers, recognizing that range anxiety re- mains a major barrier to market expansion for every player in the industry. In Europe, all manufacturers are required by law to use the same connector gauge, Tesla included.

Tesla did not invent electric vehicles, but it has launched several firsts at the right time. Retaining control of its supply chain, Tesla can be more resili- ent to potential setbacks, such as the recent chip shortage that has closed fac- tories around the world. As an arch disruptor, Tesla has challenged traditional auto business models, including reaching customers through direct sales and service rather than franchised dealerships. Customers can buy a Tesla in less than an hour from an app, rather than lose their Saturday in a protracted and often painful negotiation at a dealership.

Winning the EV Race?

The transformation of the public image of the electric car wasn't solely down to Tesla. The Nissan LEAF (Leading Environmentally-friendly Affordable Family) car, which arrived in 2010 as the world's first 100 percent electric car for the mass market, also helped to transform the otherworldly image of electric mobility. By persevering with the LEAF, Nissan demonstrated its Force-for-Good credentials. Fun and perfectly adapted for urban life, the LEAF enjoyed huge success in the 2010s as the family's second car that soon became its first. Over 600,000 were sold worldwide over the next decade.

When Tesla and the LEAF entered the market, the goal was to make EVs viable. They achieved that feat. The industry wouldn't be where it is today without those trailblazers. In 18 European countries, more than one in five new cars is electric.[36] Analysts forecast that the total global sales of electric vehicles could rise to two-thirds of all new car sales by 2030.[37]

Now the race for long-term leadership is on, as heritage car manufacturers—and newcomers emboldened by Tesla's success, like Rivian, Nio, and Lucid—respond to growing demand. It is not known whether Tesla will win that particular race. Although it remains the runaway leader in the U.S. EV market, its long-term position is far from assured, not least with the high volumes of new models in development by the Big Three and international manufacturers. However, Tesla is far from doomed. It has long since consolidated its position as a pioneer of EV development. It may become the "Intel Inside" of motoring by licensing its Autopilot-assisted driving technology, charging infrastructure, and battery production to rivals.[38] Imagine a Ford, Chevrolet, or Chrysler with a small T badge on the tailgate.

As EV adoption grows, the specialized batteries and computer operating systems EVs require may become commoditized, putting Tesla in an advantageous position given its lower production costs and superior over-the-air update technology.

Tesla has also created new markets and product line extensions. By manufacturing its own batteries in a joint venture with Panasonic, the company is expanding into energy storage systems for homes and businesses. And if cars go down a different route—say, via hydrogen or nuclear power—then Tesla may be best positioned to pivot and take its tribe of first adopters on a new journey. An Air Tesla, perhaps?

Measuring the Tesla Effect

Every company story has its setbacks, its missteps, and its limitations. While Tesla has fared well on environmental issues, the company has been criticized for people-related issues following accusations of labor and employment violations at one of its so-called gigafactories.[39] CEO Elon Musk himself has become a polarizing figure on social media and the political stage.

Tesla is primarily a U.S. success story, and the U.S. still has an overwhelmingly gas-powered market; the EV share is only 8.7 percent.[40] Compared with China, which has over 20 million battery and plug-in hybrid EVs on the road, and Europe with 12 million, the U.S. is lagging, with around seven million. Tesla's dominant U.S. position has not yet been replicated globally.

Nonetheless, Tesla is synonymous with the success of EVs. It's hard to imagine that the EV world would be where it is today without its contribution. How many business founders get to see their legacy taking shape before their eyes? Tesla foresaw the cultural change toward electric vehicles and the potential demand in the market, then got way ahead of it.

Is Tesla the New Ford?

There are still plenty of opportunities for EV companies to innovate. Batteries are emissions- and water-intensive to manufacture and recycle. Metals such as lithium, nickel, and cobalt, as well as rare earths, are in limited supply and carry ethical and environmental risks. The expense of batteries is a major reason that EVs are still beyond the purchasing power of most families.

Tesla's ambition is to keep selling cars until the last gas-powered vehicle has gone. It's a bold statement, given the 250 million existing fossil-fuel cars and light trucks on the road. Replacing them will take decades at the current rate, and even if the pace of the transition increased, could that kind of demand for batteries ever be met? In the years ahead, the ingredients needed to manufacture them may become more sought after than oil.

Other significant barriers to EV adoption remain, including building the necessary 50,000+ new publicly available charging stations required every year across the U.S. to reach the government target of 500,000 by 2030.[41] Affordably priced commercial vans and heavy-duty vehicles powered by electricity are also needed.

Then there are EVs' own environmental impacts to consider. While 99 percent of lead-acid batteries are recycled in the U.S., only 5 percent of lithium-ion batteries currently are.[42] Giving EV batteries a second life in storage could provide an alternative to recycling. While an EV battery below 80 percent of capacity will reduce the range of the car, this is not a constraint for stationary storage and could offer over a decade of additional use. Indeed, GM has designed its battery packs with second-life use in mind.[43]

Is Tesla the new Ford, representing the greatest force for innovation in the auto industry since Henry Ford released the Model T in 1908? The comparisons are tempting. Just as Ford invested in its own steel plants and rubber plantations, Tesla has created a largely unified supply chain. Just as Ford built its business around the Model T, Tesla puts most of its efforts into its highest sellers, the Model 3 sedan and Model Y SUV.

However, while Tesla is disrupting the auto market, it can't yet compare to Ford's achievement in disrupting the entire transportation system and manufacturing industry. So far, Tesla represents an evolution of the car rather than a revolution in our way of life.

Indeed, it may be that the glamor of Tesla and the EV revolution risks blinding society to the transformative innovation that it now needs. In cities especially, mass-transit systems like mobility-as-a-service, car sharing, and rental schemes, shorter commutes and mixed-purpose communities, and improved human-focused urban design could offer greater environmental and social benefits. There is still room for other transportation pioneers to move in. As we have seen in other cases, innovation creates opportunities for even more innovation.

Fueling Today and Tomorrow

Automakers need to balance the long-term need for mobility with the rising worries among stakeholders regarding climate change. Decades from now, we may live in a world where planes, trains, automobiles, and ships are all powered by nuclear and hydrogen energy. Despite public concerns around both technologies, they appear to represent the cleanest and most affordable options on the table. Yet we also know that they won't be available for several decades. So what do we do in the meantime?

The first step is accepting that, for the foreseeable future, our energy supply for transportation will need to come from a mix of sources, rather than a winner-takes-all scenario. Batteries could own short city journeys, while fuel cells and synthetic fossil fuels take charge of longer distances. An abrupt break from gasoline and diesel could prove catastrophic on a societal level. In developing countries, surely the pragmatic, inclusive approach is to provide diesel and natural gas tractors so that communities can become food-independent. That would prove a greater Force for Good than insisting on renewable energies that developing economies cannot afford. Otherwise, we risk leaving a huge part of the planet behind.

Even in developed nations, the transition to electricity cannot happen overnight. Zero-emission vehicles are coming, but not yet. In the U.S., over 95 percent of cars run on oil-based fuels. A 10-20 percent growth in EVs over the next decade would represent a major transition, but it would still leave a significant reliance on gasoline and diesel.

Besides, is the U.S. culturally ready to move from the roar of a V8 engine to the soft hum of electric? Will Americans forego private cars in favor of public transport or ride-sharing? Only when cultural shifts like these become more popular will business and government provide the mass innovation and infrastructure needed for a true mobility revolution.

Accelerate with Data

The transition to greener energy for mobility will require vast investments in R&D to make possible new infrastructure for batteries, hydrogen energy solutions, and the new architecture and operating systems needed to connect automotive and IT ecosystems. The pace and nature of the innovations we develop will decide the market shares enjoyed by various technology solutions.

Auto makers are under increasing pressure from investors, public opinion, and government regulation to publish a more detailed account of their carbon footprint, beyond mere tailpipe emissions. Manufacturing, operations, and activity throughout the supply chain all contribute to the overall picture. In Europe in particular, the rise of electric mobility and the phasing out of combustion engines is a regular topic of discussion socially and in the media. At trade conventions, it's the electric and hydrogen prototypes that feature most prominently, giving a clear indication of the direction of change. The conversation has shifted from the *why* to the *how*.

With so many materials and parts in a modern vehicle, the accurate assessment of life-cycle emissions has become a new discipline over the last 30 years. Auto makers are now asking questions like, "If we source fenders from a supplier that uses hydrogen-powered electricity, will it will drop our emissions for that part by 20 percent or more? If so, let's go with that supplier."

Manufacturers are performing extensive life-cycle assessments (LCAs) to shine a light on every activity within their supply chain. These LCAs (which I will explore in greater detail in chapter seven) can have their biggest influence in the design stages, helping engineers select materials and technologies that will drive down emissions. LCAs used to focus on fuel consumption and efficiency, due to a nagging fear among car makers about another fuel crisis akin to that seen in the 1970s. Today, that groundwork is being widened to include the whole supply chain.

Suppliers are under equal amounts of pressure to analyze and decarbonize their materials and production techniques, as a means of retaining the business of the original equipment makers (OEMs). They too will pass their demands down the line to raw materials producers. All these companies need the numbers to prove their claims about sustainability. Data helps mobility companies tell their stories, counteracting the polarizing impact of social media. Data allows companies to say, "This is why we're making this decision—and look what happens when we don't."

Of course, data collection needs to go beyond carbon accounting. What is the environmental impact of the chemicals auto makers use? What are the levels of safety in the workplace? Making the best electric vehicles in the world doesn't matter if you are polluting the atmosphere and rivers or mistreating your staff. A full automotive LCA also needs to reflect driving performance and safety requirements, as they are equally important. Companies need to take a holistic approach. Data collection, analysis, and visualization are therefore becoming a Force for Good in the auto industry.

Winning the Race

"The desire to fly is an idea handed down to us by our ancestors who . . . looked enviously on the birds soaring freely through space . . . on the infinite highway of the air." This observation by Wilbur Wright explains the motivation behind the years of creative work that Wilbur and his brother Orville invested in making the dream of heavier-than-air flight into a reality.

Like the railroads and automobiles, manned flight has made an indelible mark on society. Aviation helps to drive the development of the modern world, powering trade and tourism and connecting communities. As a Force for Good, the sector generates economic growth, providing jobs and improving living standards. Planes and helicopters offer a lifeline to those living in remote locations, aid to those in poverty, and rapid response when disasters occur.

From the Greek myth of Icarus, who soared too close to the sun, to the sketches of Leonardo da Vinci, humans have always fantasized about looking down on the land from the sky. Around 900 years before the Wrights' maiden flight in 1908, the intrepid English monk Eilmer of Malmesbury swallow-dived from his abbey spire in a pair of homemade wings. He apparently glided for a furlong (656 feet) over the river Avon, before crashing and breaking both his legs. Eilmer was lame for the rest of his life, but his only regret was "that he had forgotten to provide himself with a tail."[44]

Many other valiant inventors similarly discovered that the power-to-weight ratio of human beings—unlike that of birds—is simply insufficient for sustained flight. Lighter-than-air balloons, such as those flown by the Montgolfier brothers in 18th-century France, made flying possible, and they attracted vast crowds, firing the imagination of adventure writers like Jules Verne. An airship crossed the English Channel from France as early as 1785.

But heavier-than-air flying machines remained out of reach. The advent of internal combustion engines in the last quarter of the 19th century sent would-be aviators around the world into a fever pitch of competition. Who would be the first to take off and keep on going?

"Give us a motor and we will very soon give you a successful flying machine," declared the American inventor Hiram Maxim. He understood that imitating the propulsion systems of birds was a self-defeating exercise—but, like so many others, Maxim never achieved liftoff. Prominent scientists and financiers sank time and money into winning the race. Their experiments drew the finishing line closer, but most ended in bankruptcy, public humiliation, and often tragedy.

Why did two bicycle shop owners from Ohio achieve controlled, sustained flight when so many others failed? The Wrights had no family fortune, college degrees, or public status. The pair were simply "men of their time," according to the aviation historian R.G. Grant:

> Although they lived far from the centers of fashion and power, Orville and Wilbur Wright had grown up very much in touch with contemporary currents of thought and innovation. Their formative

131

years were in a time when new inventions proliferated—the telephone, automobiles, electric light, wireless telegraphy and cinema. Inventors such as Thomas Edison and Alexander Graham Bell were the heroes of the age.[45]

The brothers would soon add their names to that list. Their advantage lay in their systematic and methodical approach, their deep fraternal bond, and the serendipity of running a bicycle shop.

The Wrights loved to make lists. They identified three obstacles to powered flight: wing design, propulsion, and steering. Solve all three and the prize would be theirs. The puzzle of wing shape had largely been cracked by the German "birdman" Otto Lilienthal, who pioneered hang gliders in the 1880s. Before his final, fatal flight in 1886, Lilienthal had left the aviation community a wealth of detailed information on aerodynamics.[46]

Using their own homemade wind tunnel, the Wrights calculated the performance capabilities of various kinds of airfoils—streamlined surfaces designed to produce lift and drag—to build a glider in 1902. Slaves to the data, their experiments were meticulous to the extremes of obsession. They were closing in on the secret of flight.

The second problem, propulsion, was one of power-to-weight ratio. In recent years, electrical, hot-air, compressed-air, and even gunpowder engines had been tried in failed attempts to get their engineers off the ground. In 1896, fellow American Samuel Pierpont Langley, the Wright brothers' closest rival, had gained fame for propelling a steam-powered model aircraft across the Potomac River. The showman Langley was the antithesis of the sober Wright brothers. A decorated astrophysicist who had helped to standardize global time zones and close friend of Alexander Graham Bell, Langley's bid for primacy was backed by the prestige of the Smithsonian Institute and a $50,000 grant from the U.S. War Department—a budget fifty times greater than the Wrights'.

Langley calculated that steam power couldn't carry a human, so he began trialing large gasoline engines on ever-more-sophisticated gliders. However, size doesn't always matter. While Langley threw money, people, and headlines at the conundrum of flight, the Wrights obsessed over the minutiae. Through their precise experiments, they recognized that a four-cylinder 12.5 horsepower gas engine, weighing under 200 pounds (including fuel and coolant), would carry their bird far enough to meet the benchmark of sustained, controlled flight.[47] No more. No less. Finding the optimum shape for the propellers also required many days of computation, theory, and trial and error.

Controlled flight was the last and most perplexing challenge. Previous inventors had focused on getting their planes in the air. From the start, the Wrights also put their energies into flying the thing. And that's where their knowledge of bicycles demonstrated its worth. While most pioneers believed that a flying machine could be steered like a car, the brothers intuitively understood that it would need constant adjustments of balance across all three axes, much like a child learning to ride her first two-wheeler. The brothers therefore lay prone on their gliders to develop a feel for flying.

Finally, like the ancient skygazers, their big breakthrough came from bird watching. As Wilbur studied a buzzard cruising on air currents, he became transfixed by the ripple of wind on the wing tips, which appeared to give the bird lateral balance.[48] Twisting a strip of cardboard in opposite directions, he modeled a warping effect that would solve the problem of roll. Now the wings would increase lift on one side and decrease it on the other as the plane turned. A rudder counterbalanced drag, while a horizontal "elevator" at the front controlled pitch.

By the second half of 1903, the brothers were ready to fly. But so was Langley—and he took his chance first. In front of throngs of reporters and photographers, Langley launched his Great Aerodrome off a houseboat on the Potomac. It nosedived into the water.

"Langley's designs were inherently flawed," science writer David Kindy explains.[49] "While he had made limited strides in the understanding of lift, thrust and drag, he failed to see that his models when scaled up to include a human and larger engine were structurally and aerodynamically unsound, and were not capable of flight."

The ensuing ridicule ended Langley's career as an aviator and gave the floor to the Wrights. On the 17th of December 1903, the Wright Flyer took off from a wooden rail on the sands of Kill Devil Hills, North Carolina. It was Or-ville's turn to pilot the machine. He recalled: "It was only a flight of 12 sec-onds. And it was an uncertain, wavy, creeping sort of flight . . . but it was a real flight at last and not a glide."

Any doubt about the validity of this attempt was settled on their fourth attempt, when Wilbur flew the Flyer for a full 59 seconds, covering 852 feet, before crashing into the sand.

Sky's the Limit

The Wrights' success sparked an aviation frenzy throughout the U.S. and Eu-rope. Soon, new feats and records were posted, as crowds and dignitaries flocked to watch the wonderful men—and women—in their flying machines. By the First World War, technology had improved so quickly as to allow aerial bombing and dogfights. It never takes long for humans to adapt technology for warfare.

As train routes struggled to reopen after the war, the first air passenger services were formed to help fill the void, including the Dutch airline KLM, the world's oldest airline in continuous operation. In the U.S., the development of the airborne mail service greatly advanced the field of avionics (aviation elec-tronics) and brought Americans even closer together, shrinking the time / dis-tance ratio once again. Business correspondence from Los Angeles could now reach New York in two days rather than five by rail, changing the playing field for financial institutions, among others. Municipal airports with paved, rather than grass, runways were built in all major cities. The rise of the airplane (in step with the automobile) further hastened the decline of the railroads.

The public was enthralled by daring flights of adventure by pilots in re-sponse to high-profile competitions initiated by newspapers, institutions, and governments. In 1927, Charles Lindbergh flew his *Spirit of St. Louis* from New York to Paris, completing the first solo, non-stop, transatlantic air cross-ing. Amelia Earhart followed in 1932, inspiring women of all ages around the world.

While European governments chose to centralize and subsidize national flag carriers, the U.S. invited commercial operators to develop civil aviation. This arena for aggressive competition greatly accelerated innovation in technology and aircraft performance.[50]

Designers trialed groundbreaking piston engines and high-octane fuels to reach new speeds, heights, and levels of maneuverability. Manufacturers such as Boeing, Douglas, and McDonnell grew to dominate in the U.S. before setting their sights on the global market with innovations such as cabin pressurization and tricycle landing gear. Alongside airplanes, vast diesel-powered hydrogen-filled airships, many built by the German firm Zeppelin, competed on the North Atlantic routes until the *Hindenburg* went up in flames over New Jersey in 1937, with the loss of 36 lives.

After the Second World War, the jet engine would revolutionize and then democratize passenger flight, taking performance beyond the limits of piston engines. Passenger flights gradually became faster and more efficient, airliners grew larger and more durable, and services became more affordable and convenient. Domestic and international business trips became commonplace. Long-distance winter and summer holidays became a fixture for middle-class families.

Within just a few decades, almost every city on the globe was just a few days' travel away. "The modern airplane creates a new geographical dimension. . . . There are no distant places any longer: the world is small and the world is one," wrote executive and politician Wendell Willkie in his 1943 book *One World*.[51]

Supersonic Aspirations

The British and the French haven't always gotten along as neighbors. The early days of aviation brought their rivalry into the skies, as adventurers raced to become the first to cross the English Channel in an airplane (the French won). In the 1960s, however, the competition was shelved as the two governments entered into an agreement to build a supersonic passenger jet, aptly named the Concorde. Carrying just 100 people, the hook-nosed flyer aimed to cross the Atlantic in just three and a half hours, at Mach 2 speeds of around 1,350 miles per hour and a height of 11 miles.[52]

But by 1976, when the first scheduled Concorde flights, operated by Air France and British Airways, took off, the appetite for such a visionary piece of engineering had waned. Many airports were spooked by the potential for noise and fume pollution. The 1974 oil crisis turned the Concorde into a once-in-a-lifetime extravagance (return tickets from Paris to New York cost $6,000 in 2000), rather than a vision of the future. The development costs of the Concorde were written off by both governments as a national success story of engineering, but one with no apparent commercial value. Just 14 Concorde jets went into service.

Developed in the 1960s, the Boeing 2707 aimed to go faster, reaching Mach 3 speeds of 2,000 miles per hour (about the same as a rifle bullet). But the public mood changed after the plans were drawn up. Such prestigious government-backed projects had once stirred national pride, but they were deemed excessive by the end of the decade. Supersonic flight was just too loud, while environmental complaints were now growing in volume. The 2707 was mothballed in 1971.

Boeing remained ready to push new boundaries, gambling its existence on an airliner that was far beyond the imagination of even the Wright brothers. The 747 "jumbo jet," built in partnership with Pan Am, would carry twice as many passengers as its closest peer. Interestingly, Boeing's CEO Bill Allen believed the 747 would ultimately serve as a freight plane when supersonic flight became the norm.

Like the Concorde, the development was another example of expertise, technology, and data working in harmony. In a Seattle plant that was then the world's largest building—the size of 40 football fields—Boeing concentrated on the nose and wings, while subcontracting other parts to specialist firms around the country. Everything was bigger on the 747. The craft measured 195 feet in wingspan and 231 feet nose to tail, and the technological challenges in reducing its weight to a mere 360 tons were immense. The logistics needed to support and coordinate these airborne villages, from catering, runways, and ticketing strategies to hotels, check-in, and baggage reclaim, all demanded creative innovations.

Despite skepticism at the time, Allen was not the next Langley, nor was the 747 another version of the ill-fated Concorde. The 747 took off commercially in 1970—and it hasn't stopped flying since. Boeing would later consolidate its position as the dominant U.S. manufacturer by acquiring McDonnell Douglas in 1997. There have been new models of the 747 along the way that include improvements to fuel efficiency, safety, and security. The European conglomerate Airbus has provided stiff competition with its economical A380 widebody airliner. Boeing finally ceased production of its 747 in 2022, but around 350 are still in operation around the world.[53]

Addressing Demand and Sustainability

Despite the dominance of the 747 over the last 50 years, the passenger airline industry hasn't entirely stood still. New technologies have brought improvements across almost every element of manufacturing, operations, and customer experience. Avionics have continued to grow in sophistication across navigation, instrumentation, communication, safety, and landing assistance. And the pace of change has greatly picked up in recent years. The industry is currently looking to navigate two connected areas of turbulence: growing demand and climate change.

Globally, the number of flights rose annually from 24 million in 2004 to nearly 40 million in 2020.[54] The Covid-19 pandemic grounded fleets, more than halving traffic in 2020 and 2021 and causing losses of $168 billion in 2020 alone (down 40% year on year).[55] However, global demand has since quickly rebounded, with 2024 numbers estimated at 41 million flights carrying some 4.7 billion passengers globally—a 4.5 percent growth over pre-Covid-19 figures.[56]

By 2035, 62 percent of the global population will live in cities, and the number of major aviation hubs will rise from 55 to 93, due to the rising economic strength of Asia and the Middle East.[57] This demand will help drive long-term annual growth in the aviation industry of around 4 percent, requiring the global airliner fleet to grow to more than 38,000 planes with more than 100 seats by 2032.[58]

For manufacturers, the practicalities of meeting demand and honoring contracts with airline companies are made even more challenging by the second consideration: climate change. The industry has set the target of reducing aviation's net CO_2 emissions to half of their 2005 level by 2050.[59]

How do we balance the value of aviation to the world economy and our way of life with its impact on the environment?

Making efficiencies and reducing emissions while building more planes and servicing more passengers sounds like a paradox. Yet it's one that the aviation industry will need to solve. Historically, customers have proved slow to factor greenness into their booking choices, but that trend seems set to change. "Most passengers understand that aviation has a significant impact on the environment," according to a survey by McKinsey.[60] Over half of the survey respondents said they feel "really worried about climate change" and that "aviation should become carbon neutral in the future."

The report also found that "almost 40% of travelers globally are now willing to pay at least 2% more for carbon-neutral tickets, or about $20 more for a $1,000 round trip." Younger people were more willing to pay extra for less emissions.

Achieving this mark will require all the expertise, technology, and data that the industry can muster. Finding good people is a frontline priority. There is already a growing deficit in the supply of professionals in highly skilled core aviation roles such as pilots, design and development engineers, air traffic controllers, maintenance technicians, and operations specialists.[61] Manufacturers and airlines must also look outside of roles that entail traditional skills to those that specialize in different areas, such as software and hardware engineers, data and analytics experts, and AI and blockchain specialists—which puts them in competition with other business sectors. And as with those other sectors, younger digital-native recruits are more likely to work for companies that score higher in the environment and social columns.

How to Decarbonize Aviation?

The International Air Transport Association (IATA), which represents airlines, has set out a strategy for aviation to meet its emissions targets. At the top of the list is the determination that "the development of new, more efficient aircraft and engines can substantially decrease CO_2 emissions."[62] According to Boeing, sustainable aviation fuels (SAFs), which are already being used on certain commercial flights, have the potential to decarbonize aviation over the next 20-30 years and cut emissions by up to 80 percent.[63] Lighter, stronger materials such as carbon fiber and biocomposites using bamboo derivatives are being developed to reduce fuel consumption.[64]

What about the existing fleet? Operational measures such as identifying weight savings will allow aircraft to burn less fuel. According to IATA, "airlines have been investing in lightweight seats and cabin equipment and even replacing heavy pilot manuals with tablet computers. Other operational measures include single-engine taxiing, idle reverse thrust and air traffic control procedures such as continuous descents into airports and traffic-flow management that prevent unnecessary airborne holding."[65] Improvements in revenue management have meant that aircraft fly with more of their seats filled, dropping the overall carbon emissions per passenger.

Infrastructure changes are also critical, with IATA calling for "navigational improvements, making better use of airspace and streamlining the routes taken by aircraft to cut down on flight time, in addition to optimizing airport layout to improve throughput and prevent unnecessary holding."[66]

Interestingly, the IATA recognizes that a global, market-based approach is needed to address the emissions challenge until government regulations come into force. Achieving this ambitious goal will require "continued investment in new technologies and strong support mechanisms for the deployment of SAFs."

The good news is that the aviation industry has already gained momentum in recent decades. Boeing has invested "about $55 billion over the last 10 years to improve the sustainable product life cycle."[67] Its newest airplanes "are 20-30% more efficient than the in-service airplanes they replace."

Boeing's airplane designs enable parts disassembly and materials recovery of up to 90 percent of currently retiring aircraft, the manufacturer says, while it has committed to delivering 100 percent SAF-powered airplanes by 2030.[68]

Despite these important sustainability initiatives, Boeing today faces serious business challenges. Safety issues at the company made worldwide headlines in the wake of crashes involving Boeing's 737 Max airliner during 2018-2019. Thanks to these and other, less-serious incidents, the company's rating on issues such as "trust," "citizenship," and "character" took a hit in the annual Axios Harris poll on corporate reputations.[69] The company is pushing hard to earn back public trust on these fronts while continuing to set an example of environmental responsibility.

Data Makes the Difference

Boeing's digital aviation solutions, including optimized flight planning, real-time weather and traffic information, and data analytics, all support airlines in their efforts to improve flight and fuel efficiencies. For example, FliteDeck Advisor is a digital tool "that enables flight crews to make real-time adjustments to their airspeed to optimize fuel use and minimize the carbon footprint of each flight. Customers have seen 1-2% fuel and emissions reduction at cruise," says the manufacturer.[70]

Airbus, Boeing's major rival in aircraft manufacturing, has also outlined a sustainability strategy to guide the way the company does business and how it designs products and services as part of its purpose to pioneer sustainable aerospace for a safe and united world. A major focus is reducing "the CO_2 emissions of aircraft, helicopters, satellites and launch vehicles, in addition to its industrial environmental footprint at sites worldwide and throughout its supply chain."[71] Airbus's goal is to bring the world's first hydrogen-powered commercial aircraft to market sometime in the 2040s.[72]

"Today, we know our love of air travel also comes at a cost: the aviation industry represents approximately 2.5% of global human-induced CO_2 emissions," says the manufacturer. "But aviation is not the problem.

Emissions are the problem. Our approach is not only ambitious, but rather, a seismic shift for our industry."[73]

Airbus has signed on for carbon-neutral growth, which will mitigate CO_2 emissions even as air travel increases. It is also aiming for net-zero CO_2 emissions by 2050 in line with the 1.5°C goal for limiting climate change set by the international Paris Agreement. Today, all Airbus aircraft and helicopters are certified to operate on a maximum 50 percent blend of SAFs and conventional fuel. By 2030, all will be capable of flying with up to 100 percent SAF.[74]

Achieving these goals is no easy task. Airbus relies on parts, components, systems, and services from approximately 8,000 direct and 18,000 indirect suppliers from more than 100 countries. "This vast, global supplier network makes major contributions to value creation, economic prosperity and sustainable development in the communities in which they operate," adds Airbus.[75] It's no leap to say that the company's overall sustainability performance is heavily impacted by the activity of its suppliers.

As large communities in their own right, airports need to make a difference to the aviation footprint. Worldwide, more than 500 airports are part of the Airport Carbon Accreditation Scheme, taking steps such as providing SAFs to airlines, prioritizing low-energy designs for construction, and powering vehicles such as transit buses with renewable fuels.[76]

Just as with other forms of mobility, any plan to reduce aviation emissions is undermined without the means of calculating its overall impact. Tracking performance throughout the whole life cycle is necessary to show whether manufacturers and airlines are meeting their commitments—and also sends a message to environmentally conscious customers. Airlines are increasingly giving travelers greater choice over their own carbon footprint by sharing information about the age of planes at the point of sale. Green planes will need to offer the same levels of convenience, safety, and comfort to passengers as traditional aircraft.

In aviation, major steps forward in the race to reduce carbon emissions will come from greener aircraft technology such as electric and hydrogen-powered planes. Full life cycle analysis likewise enables reductions in carbon emissions in the manufacturing of aircraft and production of SAFs.

Because fuel is a major cost for an airline, any innovations that can reduce carbon emissions through fuel efficiency are good for business too. By lowering passenger emissions, airlines can lower fares and so grow faster. However, while work is underway to reinvent aviation as a sustainable form of travel, so much still needs to happen if it is to meet the ambitious targets of the future.

A successful green transition for air travel will require the mobilization of the entire ecosystem. Like the Wrights in their bicycle shed, today's engineers must find new formulas for propulsion, aerodynamics, materials, and power-to-weight ratios. They must think big and small, looking to the world around them for inspiration and searching for progress in the details.

The prize for getting it right first is every bit as tantalizing as the age-old goal the Wright brothers first attained.

Trade Winds of Change

Since humans first took to sea in dugouts and rafts around 10,000 years ago, innovations in propulsion, navigation, and cargo have pushed civilization forward. For the last 100 years, marine propulsion has been dominated by diesel-powered engines. Navigation is now directed by satellites, with just-in-time logistics ensuring an optimum rate of progress to port in largely automated ships.

When Malcolm McLean died in 2001, the father of containerization was applauded by *Forbes* magazine as "one of the few men who changed the world." His brainchild—standardized steel containers that could be stacked on boats, carried on trains, or hauled by trucks—has made shipping faster and easier, contributing to higher living standards in every country in the world. As with Bohlin's seat belts, McLean's good idea has touched the lives of billions.

Today, 90 percent of world trade is transported by sea on vast ships, and this figure is only likely to grow.[77] Industrial commodities such as fuel, iron ore, metals, and feedstock are almost entirely reliant on transport by sea. The market for affordable food and manufactured goods—the very lifeblood of the international economy—would wither without a robust shipping industry.

However, this progress has come at a cost. Shipping consumes more fuel than any other mode of transport and produces around 3 percent of global greenhouse gases.[78]

In all, fuel accounts for 60 percent of a ship's lifetime costs, which therefore offers a major opportunity for reducing CO_2 in line with the International Maritime Organization's (IMO) target of net-zero GHG emissions from international shipping by 2050. The IMO is also committed to ensuring "an uptake of alternative zero and near-zero GHG fuels by 2030."[79] Hitting those marks won't be easy. Replacing hydrocarbons with clean energy sources has been compared with the shift from sail to steam in the second half of the 19th century.[80]

The deterrents and the incentives are both meaningful. Companies that fail to meet carbon reduction milestones risk losing their license to operate. For innovators, the green shipping transition will need upwards of $1 trillion dollars by 2050, offering a major inducement.[81] According to the U.K. government, the global market for maritime emission-reduction technologies could reach $15 billion per year by 2050.[82] With lifespans of 30 years, many ships in the world's 100,000-strong merchant fleet will need to be retrofitted with greener engines over the next two decades.

The best choice for fuel is still in the balance, however, and Maersk, the world's largest shipping firm, is determined to be the first to work it out. "The science is clear," said Maersk CEO Vincent Clerk. "We must make an impact in this decade, and we are now accelerating our climate ambitions by ten years and committing to be net zero across our business and value chain by 2040 with 100% green solutions for our customers."[83]

Transitional fuels like natural gas and blue hydrogen, whereby the carbon emissions from burning natural gas are captured and stored, may offer a viable interim option. Biofuels derived from waste feedstocks have significant potential, although they face problems in the supply of biomass. Zero-carbon ammonia, while currently unstable, presents a long-term alternative as storage technology improves. Green methanol, produced from sustainable biomass and renewable electricity, is emerging as an early front-runner, despite a current lack of bunkering infrastructure in ports.

In 2020, the IMO also introduced a new limit of 0.50 percent m / m (mass by mass) on the sulfur content in the fuel oil used onboard ships to "improve air quality, preserve the environment and protect human health," marking a significant reduction from the previous limit of 3.5 percent.[84]

The sector has shown its ability to harness the Formula for Good—expertise, technology, and data—in vessel management solutions that help operators plot journeys with optimal fuel consumption.

Shippers increasingly rely on real-time data to help minimize risks such as delays, empty and thus profitless journeys, bad weather, and adverse currents. These new data-powered ship plans offer a level of functionality that leaves standard spreadsheets trailing in their wash.

Perhaps the ultimate solution to green shipping operations is still to be found. Ship design and operating procedures may need to change dramatically in the future. Competition between major ports and the world's busiest shipping lanes, such as the Panama and Suez Canals, the Bosporus, and the Straits of Hormuz and Malacca, will intensify. There are choppy waters ahead. But don't underestimate the potential of businesses to innovate. In 50 years, we may see self-fueling ships that scoop salt water from the oceans and convert it onboard into green hydrogen. How good would that be?

The Rise of Smart Cities

Today, more than half of all people in the world live in cities, a proportion that's expected to reach two-thirds by 2050. Maintaining environmental, social, and economic sustainability is therefore a global issue, with mobility at the heart of progress.

A smart city uses digital solutions like IoT and wireless networks to make traditional services—including those related to mobility—more efficient, bringing greater benefits to its citizens and businesses. Smarter urban transport networks that manage resources and emissions are top priorities, alongside better water supply and waste disposal facilities, more efficient energy production to light and heat buildings, and safer streets and more equitable living conditions for vulnerable sections of the population. Improvements in IT can therefore improve quality of life and save lives. By helping to "make cities inclusive, safe, resilient and sustainable," these improvements can enable governments to meet goal number 11 on the list of 17 Sustainability Development Goals adopted by all member states of the United Nations in 2015.

Currently, most cities fail to take advantage of economies of scale across their many services and districts. Smart cities enable a more integrated approach. A simple example is the use of intelligent traffic management systems to help reduce congestion, which has a major environmental, social, and economic impact. City councils are increasingly seeking new ways to shorten journey distances and remove cars from the road, particularly in downtown areas.

Smart mobility is an important element of the smart city, using wireless communications to coordinate the many different modes of transportation used by citizens. Real-time data analytics and machine learning are applied to obtain the benefits of safer and more efficient transportation systems, which can reduce levels of congestion, pollution, and road-traffic accidents.

So far, few cities have developed a smart mobility strategy to meet local challenges. The level of sophistication and integration required will vary depending on investment and capabilities. Mobility-as-a-service (MaaS) platforms are becoming more commonplace, as they allow travelers to plan, book, and pay for different types of transport on the same system. Individuals, families, and businesses may choose to pay for a subscription that removes the need for car ownership.

The rise of autonomous and connected vehicles, along with e-mobility, is sparking innovations around circulation and infrastructure, with car manufacturers devising means to monetize the usage of vehicles, rather than just the vehicle itself.

Connected Thinking

The cities that are working toward using data to operate whole cities, rather than implementing piecemeal solutions here and there, are the ones that are accelerating away from the pack.

The major challenge is less about the use of technology and data and more about how to govern those processes. The need for governance that stretches across industries and different tiers of government is often the barrier to progress. City councils need to bring the public on board and show that they are solving a local need.

North American cities such as Toronto, Chicago, and Chula Vista, California (which mobilizes a flotilla of drones as first responders to help keep the peace) are all leading proponents of smart mobility solutions. A common factor across successful regional efforts is the presence of strong connector-leaders who see the bigger picture. They know how to bring people from a wide number of organizations together and then navigate the risks, recognizing that a local government or NGO's view of risk is very different from that of a private-sector company. Of course, there is a real danger that when the strong leaders leave, a significant drop-off in activity follows.

Cities are increasingly taking a platform-based approach to support the integration of data and to better cope with the speed of technological change. Whole regions are developing data exchanges that provide an environment for sourcing, sharing, and partnering over data. For example, Data Mill North in the north of England was set up by Leeds City Council as an open data repository to catalyze innovation in the area.[85] The Amsterdam Smart City program in the Netherlands has demonstrated the shared benefits of a data vault that gives businesses and developers open access to traffic and transportation data.[86]

Worldwide, urban digital twins are becoming more popular, whereby vast areas are connected by IoT and sensors to create a real-time picture of what's going on at street level. The New South Wales Digital Twin in Australia invites the regional council, planners, and the wider community to interact with data for services from public transport, tourist attractions, parking and electric vehicle charging spaces to children's play equipment and off-leash dog parks.[87]

The city of Los Angeles aims to transform itself into a smarter metropolis by the 2028 Summer Olympic Games. This will be powered by ubiquitous 5G connectivity, internet-connected sidewalk kiosks, single-payment micro-transit options, and the rollout of 10,000 public EV chargers. The city council has revealed plans for "ethical proactive technology" that will help identify "fire, violence, or other risks to the health and safety of L.A. residents" even before a 911 call is made.[88] Perhaps if this system had been in place during the major L.A. fires of 2025, the area of damage might have been contained.

Since 2017, the IESE Cities in Motion study led by the IESE Business School in Spain has ranked the world's 183 smartest cities across the nine factors of economy, human capital, international profile, urban planning, environment, technology, governance, social cohesion, and mobility and transportation. Every year, London has come out on top, most recently followed by New York, Paris, and then Tokyo.[89]

New York was ranked top in mobility and transportation, due to the city's "highly developed subway system, with the largest number of stations. The city also has a good system for bicycle, scooter and moped rental, and ranks fifth in the number of inbound air routes," said the report's authors.

Imagining the future with a high degree of accuracy is a futile task. However, we can be confident that for the next 50, 100, or even 300 years, mobility will need to be a Force for Good. There's a temptation to merely adjust existing models of transportation, rather than search for a different model entirely. But will that slow, incremental approach meet the needs of tomorrow's culture?

As Henry Ford might have said, we don't want faster horses. We need to innovate a new form of transportation altogether. Business will find a way; I'm sure of it.

Key Takeaways

- The mobility sector has an outsized impact on the environmental and social well-being of our planet.
- The U.S. railroads created a culture of opportunity that inspired the development of the automobile and airplane—which would eventually lead to the railroads' decline.
- Today, digital innovation offers hope for the recovery of passenger rail services, as demonstrated by railways in Asia and Europe.
- Pioneers like Henry Ford and GM's Alfred Sloan grew their automotive companies into global leaders and influenced modern culture beyond recognition.
- Contemporary car manufacturers must find a strategy that balances the needs of society with environmental concerns.
- The Wright brothers won the race to fly with a combination of expertise, technology, and data, setting an example for generations of aviation innovators.
- Modern aviation leaders are under pressure to meet the twin challenges of increased demand and reduced carbon emissions.
- Shipping is also engaged in a race to find sustainable sources of fuel while harnessing data to achieve optimal journeys.
- Finally, the era of the smart city is taking root, using IoT technology to improve the quality of life for billions of citizens worldwide.

5

WHY SUSTAINABILITY BECAME A 21ST-CENTURY FORCE FOR GOOD

- Introducing the benefits of enterprise sustainability management
- The Covid-19 pandemic as a catalyst for change
- How a focus on People and Planet results in high Performance
- Materiality matters
- The critical importance of operationalizing sustainability

"Business shapes the world. It is capable of changing society in almost any way you can imagine."—Anita Roddick, founder of The Body Shop

n this chapter, I'll discuss how businesses can harness the power of sustainability to become a Force for Good in today's market. Of course, not every business will lead the way. As mentioned earlier, the Lions will pick up sustainability and run with it, while the Ostriches will bury their heads in the sand—and risk business failure.

But first, it's worth defining several terms that crop up repeatedly when talking about sustainability. Let's start with *enterprise sustainability management* (ESM). This describes how companies *operationalize* their environmental, safety, social, and governance objectives as a means of bringing greater efficiency and productivity into the day-to-day running of their business.

Regulatory and societal pressures are driving companies to apply sustainability principles to their operations. As a result, not only are companies increasing circularity and efficiency across their operations, but they are also improving and increasing the lifetime and resiliency of their assets. In this way, companies are enhancing the operational excellence of their organizations.

ESM allows Force-for-Good companies to take sustainability beyond reporting and into their product innovation, safety programs, employee relations, culture, values, and much more. It provides a sustainability tool kit to galvanize higher business performance. Rather than treat sustainability as a static outcome, it's elevated to be an integral and active part of the process. Better business performance is now the result of sustainability, rather than the other way around.

Indeed, many of the world's largest organizations are now reaping the tangible benefits of interlinking their operational and sustainability goals.[1] They are *applying* ESM to rewire their entire value chain from supply chain management to product design, operational risk management, sales and marketing, and customer relations.

When viewed in this way, sustainability is not a cost, but a source of innovation. The bottom line is that when decisions on procurement, logistics, supply chain, manufacturing, process safety and packaging are taken with sustainability in mind, those choices have a positive financial impact, too.

This synergy between sustainability and business operations is known as the 3Ps—People, Planet, Performance—whereby actions taken on behalf of People and the Planet result in optimized Performance.

I'll also use the word *materiality* to describe how companies discern which People or Planet criteria are most relevant to their specific business. There isn't a one-size-fits-all approach to sustainability. For example, a chemicals manufacturer operating a series of factories will tackle carbon emissions, workforce safety, and pollution in a different way than a large insurance company that employs several thousand employees in urban offices. What's material to one is less material to another.

Looking Beneath the Hood of Sustainability

ESM is intertwined with the workings of every business. Understanding the moving parts of sustainability helps companies to convey an even more compelling and accurate story to stakeholders such as employees, investors, the media, and the government. Let's look at how the 3Ps fit into this story.

In simple terms, People is about the human impact of your actions. The safety and fair treatment of employees are crucial concerns. Companies must also recognize their role within the communities in which they operate. Reputations are built over time, but can be broken quickly and are then hard to fix.

Planet covers the energy, water, and resources that a company uses, the waste that is produced, and its impact on the land, water, and biodiversity. Is your environmental footprint growing or receding? Is it under control? Carbon emissions and climate change are central considerations.

Importantly, the evidence shows that when companies use the opportunities of operationalized sustainability to their best advantage, they create value for all their stakeholders—and enjoy higher Performance at the same time.

Firms that take a Lion's approach to sustainability run a more efficient and cost-effective business, retain good people, and attract investment over the long term. Companies that understand which People and Planet issues most affect their business—the material impacts of sustainability—outperform those that don't. Yes, organizations really can be profitable and responsible at the same time; in fact, in today's world, doing both is a necessity.

The growth of interest in sustainability has coincided with the bulldozing of the ivory towers in which organizations formerly liked to make their decisions. Consumers, employees, and investors are keeping a closer eye on those companies that have a direct bearing on their short- and long-term futures. Social media can dismantle decades of goodwill in a weekend.

The public consensus now supports sustainability. Nine out of 10 Americans agree that there should be a standardized structure for companies to report the social and environmental impacts of their business practices—something that the European Union is already bringing into force through legislation, as we'll discuss in chapter six.[2]

How Sustainability Results in Better Business

According to analysts at McKinsey, five important reasons link sustainability directly to cash flow:[3]

1. Facilitating top-line growth
2. Reducing costs
3. Minimizing regulatory and legal interventions
4. Increasing employee productivity
5. Optimizing investment and capital expenditures

We can build on those five key reasons as follows:

1. *Growth:* Sustainability is inherently innovative. With more sustainable products, companies will attract new customers. Competitors that don't may lose out.

2. *Efficiency:* Companies need to adapt their products to address concerns around carbon emissions, energy, waste, and water consumption, for

example. These are all avenues for cost savings—or cost increases, given that unsustainable packaging and waste disposal may cost more in the future.

3. *Compliance:* Companies that stay a step ahead of environmental regulation can benefit by winning subsidies and government support. Those that fall behind can face financial penalties and sanctions.

4. *Employees:* There is a direct correlation between a business's reputation in terms of People and Planet criteria and its ability to attract and retain top talent. On the flip side, reputational damage is hard to repair.

5. *Capital:* Investors are increasingly looking for companies that meet their standards for sustainability, as the evidence shows that these are better bets in the long term. Investment funds are growing their pots for sustainability-focused capital. Conversely, a lower sustainability rating may dry up investment. Companies may find themselves in a vicious spiral whereby they struggle to find the capital they need to finance change.

Doing the Right Things at the Right Time

Sustainable businesses also rely on sound governance. This is the mesh of levers and pulleys beneath the surface that ensures a company controls itself effectively, makes sound decisions, stays compliant, and is responsible to its stakeholders. By putting the right governance in place, the business will stay ahead of problems before they occur. Companies without a robust system of guardrails and procedures can lose sight of their obligations to People and the Planet, which will eventually result in lower business Performance, too.

In the table on the next page, I've outlined the key areas of consideration for businesses when looking to implement enterprise sustainability management. Of course, these are always evolving, but these factors provide a firm foundation upon which to build one's thinking.

Technology is sometimes added as a governance criterion for ESM, as stakeholders are attracted to companies that can mitigate threats such as cyberattacks, assaults on data privacy, fake news, deep fakes, and the dark

PEOPLE	PLANET	GOVERNANCE
HEALTH & SAFETY	GREENHOUSE GAS EMISSIONS	BUSINESS ETHICS
EMPLOYEE RELATIONS	CARBON MONITORING	RISK MITIGATION
INCLUSIVE WORKPLACE CULTURE	ENERGY USAGE	REGULATORY COMPLIANCE
CUSTOMER RELATIONS	AIR EMISSIONS	TRANSPARENCY
LABOR STANDARDS	WASTE & RECYCLING	BOARD STRUCTURE
HUMAN RIGHTS	WATER MANAGEMENT	EXECUTIVE COMPENSATION
COMMUNITY RELATIONS	RAW MATERIALS	SHAREHOLDER RIGHTS
HUMAN CAPITAL MANAGEMENT	SUPPLY CHAIN MANAGEMENT	CYBER SECURITY & DATA PRIVACY

KEY ISSUES TO CONSIDER WHEN IMPLEMENTING ESM

web while taking advantage of fast-growing technologies such as AI, machine and deep learning, robotics, data mining, wearables, and nanotechnology.[4]

The Small Matter of Materiality

The need to identify material topics is an imperative for businesses and investors alike. The Sustainability Accounting Standards Board (SASB) defines material issues as "those with evidence of wide interest from a variety of user groups and evidence of financial impact." It's worth accessing the SASB Materiality Finder to look up material issues for specific companies or industries and compare industries side by side.[5]

Over the last 10 years, the Global Reporting Initiative (GRI), an independent international organization, has helped businesses and other organizations "take responsibility for their impacts by providing them with a global common language to communicate those impacts." The GRI has worked with several stakeholder groups to identify the most material sustainability issues in different sectors, which are described in the G4 Sustainability Reporting Guidelines.[6]

George Serafeim at Harvard Business School has demonstrated the link between material action and investment performance over the long term.[7] His investigations indicate that "firms with good ratings on material sustainability issues significantly outperform firms with poor ratings on these issues. In contrast, firms with good ratings on immaterial sustainability issue do not significantly outperform firms with poor ratings on the same issues." Serafeim goes on to say, "For example, managing climate change risk can be strategically important for some firms, while employee health and safety issues are more likely to be strategically important for other firms."

In their impactful 2015 report *From the Stockholder to the Stakeholder: How Sustainability Can Drive Financial Performance*, authors Gordon L. Clark, Andreas Feiner, and Michael Viehs close by saying:

> It is in the long-term self-interest of the general public, as beneficiaries of institutional investors (e.g. pension funds and insurance companies), to influence companies to produce goods and services in a responsible way. By doing so they not only generate better returns for their savings and pensions, but also contribute to preserving the world they live in for themselves and future generations.[8]

Focus on People

To operate successfully, the standard business practices of a firm need to remain acceptable to its employees, stakeholders, and the general public. This interdependence is often called a *social license* or *social contract*. Think of it as a deal with the people and community impacted by the company. As discussed in chapter two, this is a good example of the need to balance value to the stakeholder and value to the enterprise. The community benefits from jobs and economic activity. The enterprise can find a place in which to reach its full potential, often aided by tax breaks or other incentives. But if the social contract breaks down, then both the community and the enterprise suffer.

Of course, the need for organizations to look after their people and make a positive contribution to the communities in which they operate is not a new concept. So why has the focus on this connection been intensified?

Social media has made a difference, forcing companies to take a more proactive stance to avoid the stigma of being labeled irresponsible. Employee rating engines such as Glassdoor are increasingly used by candidates to check whether prospective employers share their values. Companies recognize the reputational benefits of appearing in Fortune's Best Companies to Work For and rating platforms such as Great Place to Work.

In the competition for talent, especially for young, tech-savvy workers, companies need to show that candidates will feel rewarded, both financially and emotionally. Salary is no longer the only consideration for many. They judge potential employers on their work / life balance, support services, and attitudes toward community issues.

Building a reputation as a good place to work results in better financial and competitive performance. Alex Edmans, professor of finance at the London Business School, has demonstrated the relationship between employee satisfaction and business success. "A satisfying workplace can foster job embeddedness and ensure that talented employees stay with the firm," he found.[9]

The social impacts of business supply chains have also come under scrutiny in the last decade. For example, the tragic Rana Plaza factory collapse in Bangladesh in 2013 cast a powerful light on sweatshops that produce a huge volume of fast-fashion garments worn throughout the world. While more can still be done, many retailers have since taken steps to ensure greater responsibility among board members, to enhance transparency, and to involve external auditors. This is done as a way of reassuring employees, customers, and investors that the company understands its supply chain and the importance of managing it responsibly.

In 2020, U.K. fashion leader Boohoo saw its share price crash when details emerged in newspapers about poor working conditions among its suppliers. The retailer has since taken measures to make its sourcing more transparent and sustainable, but its share price is still a fraction of the value it was before the scandal emerged.

In the last 50 years, there has been a concerted drive toward safety in heavy industries such as chemical manufacturing, and businesses are now being better regulated. Trevor Kletz was the godfather of process safety in the U.K. in the 1960s. When firms complained about the need to meet regulations, he famously retorted, "If you think safety is expensive, try having an accident."

Bringing the different disciplines of sustainability under one roof provides responsible companies with an opportunity to better communicate their hard-earned track record for safety. At the same time, they recognize that by producing less waste and emissions, they are not just being environmentally sound; they are also making products more efficiently and perhaps more profitably. The fear of future liability as a result of cutting corners is better understood today than ever before.

Gear Shift in Engagement

David Batchelor joined Sphera as chairman of the board of directors in 2022, bringing extensive international executive experience from his 40-year career in risk, insurance, and capital markets. He has charted the noticeable shift in sustainability engagement during the past few years, especially during the Covid-19 pandemic:

> In 2020, there was a different type of thinking introduced into the workplace. I'm not suggesting that sustainability didn't exist prior to that. But the pandemic certainly was an accelerator for sustainability primacy. Issues such as diversity, equity, and inclusion, climate considerations, reputation, and community had been on the

agenda for businesses for quite some time in varying degrees of profile and importance, but many hadn't made the level of progress that you would have hoped for. They are now firmly at the forefront in the thinking around strategy and purpose and board responsibility.

In Batchelor's experience, employee well-being was a particular focal point during the pandemic. Businesses realized, almost overnight, that they needed to look deeper than the nine-to-five. A lot of other aspects of workplace relations were exposed by the pandemic that are now, quite rightly, front and center. Technology utilized by business was fast-tracked by several years—and changed forever. The way that businesses operate, hybrid working, mental health, and the value of flexibility in terms of work / life balance are all important. Companies are now thinking more clearly about the overall well-being of their employees as a crucial component for success.

People also got a time-out to think about what mattered to them personally and what mattered to business. As they reconsidered their priorities, many workers looked to their employers to offer better work / life balance. "The purpose of organizations has become much sharper focused beyond just the outcomes of strategy, which generally tend to be financial metrics around performance," Batchelor believes:

> We got to spend more time with our families. We valued what was going on in the classroom, in food stores and hospitals, perhaps more than we had prior to the pandemic. Businesses were similarly thinking: How do we retain customers? How do we engage with customers when face-to-face has disappeared? There were a huge number of learnings that came out of the pandemic, although I would have preferred to have gotten this type of thinking and reevaluation by some other means than a global pandemic. But it was a real catalyst and enforced accelerant for some good things in business, helping them to surface issues that had lingered down the agenda for too long.

The next step for sustainability is finding a common trajectory. There is significant diversity regarding what businesses view as material considerations—and therefore what they should be doing in response. Not all criteria are imperative for every enterprise.

From a governance standpoint, companies need to look at sustainability within their strategy and think about how to interact in a more meaningful way with broader stakeholders, Batchelor says. Sustainability shouldn't be considered as a parallel business initiative that conveniently ticks some regulatory investor boxes; instead, it should be unbundled with the material issues affecting an organization, forming an integral part of strategy culture and purpose. "It's worth considering which other stakeholders you may not have envisaged or who aren't in your line of sight, but clearly have an impact on you—or you on them," Batchelor notes.

Finding the right data platform can help companies identify the more meaningful parts of ESM that are relevant to them. Batchelor observes:

> By gathering the necessary, quality data, companies can determine an outcome and a course of action, while avoiding greenwashing. It's important to join data with domain knowledge, rather than seeing sustainability as just a technology challenge. You need that deep-seated understanding and expertise element to determine what data you need to harvest and how it is relevant to your business and stakeholders. To my mind, that's what makes the difference between general boilerplate data and quality data that counts in terms of action and outcomes.

Anecdotally, the shift toward sustainability has been extraordinary in recent years, in Batchelor's experience. "I mean, it's hard to have a conversation without sustainability popping up, especially at a board or C-suite level!" he says. "You can sense the appetite for people, first of all, to be educated about sustainability and to understand what it is and isn't. Secondly, they want to understand how it affects their business individually. That's the missing fil-ter. What are the material issues to their business?"

Batchelor maintains that companies have achieved some great feats in the sustainability field over many years, but they haven't branded them as such. Organizations have done a lot to move the sustainability agenda forward as part of their efforts to stay competitive and relevant. The power dynamic in terms of hiring of talent has shifted from the employer to the employee, and it will no doubt switch again depending on macrocircumstances. But whatever the dominant position may be, it is always about talent competition, and em-ployees are making judgments about companies based on their purpose rather than just their financial strength and current positioning. Batchelor says:

> These things are on the minds of a lot of organizations, be they large publicly quoted global organizations or small- to mid-sized compa-nies that may be part of the supply chain of a bigger beast. Every-one's on that same journey. The smaller supply chain companies need to align with the bigger ones on diversity or on environmental concern. And larger organizations need to support and engage re-sponsibly with their supply chain and the communities they operate in.

As both a mentor and mentee of female entrepreneurs, Batchelor takes a particular interest in diversity. He has seen firsthand the commercial benefits of bringing new voices into the conversation:

> The people closer to the coalface are usually the ones who give the best recommendations in terms of improving work practices. They tend to know what customers are really saying. I've found that com-panies have taken great strides in recognizing the value of diversity as a business issue and how it contributes to success. It goes further than a tick box around the gender pay gap and gender blend. Busi-nesses have started to realize that there is true business value.

No Business Is an Island

Continuing with the People-related aspects of sustainability, David Batchelor has also witnessed an increase in community engagement—which, again, is not a new concept:

> Successful companies have understood for decades that they can't operate in isolation. They need to be part of their ecosystem. They wouldn't be successful otherwise. Customer market research and sensible strategic planning for the benefit of the organization have been going on forever, or companies wouldn't remain in business. But that level of engagement has moved up a notch. Customer engagement has increased too, in both directions, as customers are now more interested in what you are doing and how your sustainability agenda is playing out.

It clearly pays to align with the views of your community.

The societal aspect was always on the agenda in the form of corporate social responsibility or in relation to helping colleagues to support charities and feel good about their communities. Today, there's a much more deliberate approach in terms of the impact on societies and communities that businesses have. "And that gives really good feedback in terms of what's going on in the community," says Batchelor:

> Companies need to be more aware of the societies in which they operate, as that will translate into the way they conduct their business and develop products and services. This outward-facing perspective gives businesses a chance to engage with the regulatory environment in a way that perhaps they haven't engaged in the past. They can hold more pragmatic, open discussions that are more beneficial for both sides. The sustainability agenda has provided a good catalyst for stakeholder engagement. Those companies that are engaging will end up being better businesses.

During the pandemic, Batchelor notes, even though society was pulled apart, businesses found themselves *closer* to their employees:

> They had to genuinely care and be responsible. It was the same for their supply chain, because they suddenly found dependencies they didn't recognize and aspects of their supply chain under stress, which directly impacted them, too. There was a sharper recognition of who's in your supply chain and what is going on, as opposed to just extracting the maximum value that you could. This has led to more altruistic engagement, because their failure could cause your business to fail. It's a work in progress, and barriers are being erected due to antiglobalization views, but ultimately open borders managed responsibly are in the best interest of all.
>
> People and businesses were drawn out of their cocoons. There's an increased receptiveness post-pandemic. Before, people were just doing the stuff they needed to do. Now, they look up and smile. The pandemic brought two unfortunate years, but the level of a different type of engagement and recognizing what we may have previously taken for granted in society was quite unique. My hope is this continues.

Focus on Planet

For enterprise sustainability management, Planet covers energy use, water consumption and waste management, impact on resources such as land, waterways, and clean air, and biodiversity. Climate change and CO_2 emissions are topics that have a greater or lesser degree of importance for businesses, depending on their carbon footprint—although every company is expected to play its part.

The modern environmental movement in the U.S. has its roots in the 19th and early 20th centuries with the establishment of the first national parks in the U.S.—Yellowstone, Sequoia, and the Grand Canyon. In 1918, the

Save the Redwoods League had the foresight to buy the last old-growth red-wood trees. It's unlikely that Hyperion, the world's tallest tree at 379.1 feet, would still be standing today if they hadn't.

In the 1960s, the World Wildlife Fund was launched with a mission to create a world where people and wildlife could thrive together. Rachel Carson's 1962 bestseller *Silent Spring* warned of devastation from pesticides, particularly the poison DDT. DDT was banned 10 years later. In 1969, the Cuyahoga River infamously caught on fire, drawing attention to the high levels of industrial pollution.

Major legislation in the 1970s sought to protect endangered species and marine mammals, bringing tighter regulation on industrial effluence and waste dumping. In the 1980s, Greenpeace campaigned to Save the Whales, taking on the might of international whaling fleets in its schooner, the *Rainbow Warrior*.

Spokespeople emerged in business to voice the concerns of their customers. In the U.K., The Body Shop launched its cosmetics, skincare, and perfume chain in 1976 "with a belief in something revolutionary: that business could be a force for good." Founder Anita Roddick became a pioneer of "cruelty-free" products that empowered women and gave a fair deal to farmers and their communities.

"Business shapes the world. It is capable of changing society in almost any way you can imagine," Roddick said. "Social and environmental dimensions are woven into the fabric of the company itself. They are neither first nor last among our objectives, but an ongoing part of everything we do."[10]

Many mission-driven beauty companies have since followed in Roddick's footsteps, often reinvesting their profits in social causes. Major beauty brands today recognize consumer expectations to manage their impact, including the impact of their supply chains.

Innovation as Environmental Action

Climate change and the need to address sustainability regulations have risen up the boardroom agenda. According to Hari Osofsky, dean and Myra and James Bradwell professor of law at the Northwestern Pritzker School of Law, who has published extensively on climate change law, there are two drivers in play here—risk and opportunity:

> Corporations know enough about climate change science to recognize that climate change is real, so there is a real need to adapt and innovate. For big infrastructure projects, say, it pays to take an interest in what the regulations are likely to be in the future. CEOs don't want to set themselves on a corporate path that will turn into a financial dead-end due to regulation.

On the other hand, leading corporations will see a changing regulatory environment as an opportunity for differentiation. By getting out ahead of regulation through innovation, companies are in a stronger position when the legislation arrives. They're not the ones playing catch-up.

We see the impact of this surge in environmental awareness throughout our daily lives. Huge numbers of product innovations are going to market as brands respond to the public's desire for greener, healthier, happier choices. Organic, plant-based, and vegan alternatives are rapidly increasing their square footage in supermarket aisles. Manufacturers are finding ways to reduce food miles through local sourcing. Grocers and confectioners are also taking steps to reduce the fat, salt, and sugar content of their products in line with social demands on health.

Twentieth-century innovations such as packaging and refrigeration, which have made such a profound difference to quality of living in the last 100 years, have needed to recenter their efforts in response to concerns around pollution, emissions, and waste. Manufacturers are innovating once again to increase recycling and accelerate the circular economy. Those who fail to adapt will eventually find themselves squeezed out by consumer action or regulation—or both.

The disposable diaper company Pampers, owned by Procter & Gamble, was a success story of the postwar baby boom in the 1950s, offering a convenient and affordable alternative to the time-consuming cloth diaper. The story goes that P&G engineer Vic Mills got so fed up with cleaning and grappling with his grandson's cloth diapers that he set about designing something simpler.[11] Disposables already existed, but P&G used its second-mover advantage to change the sector forever.

Pampers rose to become P&G's first brand to make $10 billion in annual sales, keeping 25 million babies dry in over 100 countries. The numbers are huge, especially once you factor in that children might use several thousand diapers until they are potty-trained, as well as tens of thousands of baby wipes. Most diapers contain plastics that can break down into microplastics and chemicals that can leach out into waterways, damaging biodiversity.[12] In 2017, a 250-meter, 130-ton "fatberg" made of flushed baby wipes and grease was discovered clogging the sewers of London.[13]

Pampers now states that its biggest opportunity to reduce environmental impact is to focus on reducing materials. The manufacturer has innovated technological solutions to "decrease the weight and packaging of diapers by almost 50% over the past two decades." Its diapers now use substantially less material compared to six years ago, which, it claims, "equates to 750 fewer nappies going into the bin during the approximate period of time in which a baby is in nappies."[14] Ten of its factories are "operating at zero waste to landfill, which means that nothing that enters those factories is wasted," with the aim that every Pampers site will follow suit.[15]

Investing in the Energy Transition

As one of the world's largest investment firms, Blackstone Inc. believes that accelerating decarbonization, creating inclusive opportunities, and building strong foundational governance are crucial to developing resilient companies and assets that deliver long-term value. Managing director Elizabeth Lewis brings a wealth of experience in these areas, drawing on her experiences in energy transition and climate solutions over the past two decades.

"The job I have now didn't exist just a few years ago," she says. "The market is very different from when I started out."

Lewis identifies several reasons for the transformation. First, there is more and more evidence, both anecdotal and academic, which shows that, across sectors, taking factors such as climate risk, emissions reduction, and diversity of workforce into account when making investment decisions leads to lower risk and better investment performance.

The second game-changer was the Covid-19 pandemic, which made people stop and think about leadership, society, inequities, and healthcare, in addition to labor and workforce issues. "Investors are realizing they can play a more active role in providing solutions to these challenges," says Lewis, "and there are really good business opportunities available."

Lewis offers several pieces of advice for companies:

> The first step is thinking with a long-term mindset. Try and get out of the quarter-to-quarter vortex. Consider the investments you can make in human capital, hiring people who have knowledge of emerging climate technologies, or with data expertise in energy transition.
>
> When you're writing the budget for this year, what capex [capital expenditure] investments can you schedule for next year, which will have longer-term payoffs? These will help you develop a new, cleaner product and lower your emissions, while gaining access to new customers who are climate-friendly.

By developing a road map, companies can then budget accordingly, and subsequently have a longer-term perspective. Lewis continues, "The key is getting started. You don't need to have it all figured out immediately. By making small steps and really thinking about building out your business, it can pay off. Seek help."

The right mindset is everything. "It's important to spot and seize the opportunities to drive value," she concludes.

Setting Strong Foundations for ESM

A sustainable business needs to have the right governance, systems, procedures, and controls in place to make the right decisions—"right" as in commercially successful while also acting in accordance with community values. Governance helps to ensure that these two "rights" keep away the wrong.

While they weren't always named as such, concerns around managerial accountability, board structure, and shareholder rights date back over several centuries. The issue of governance grew during the 16th and 17th centuries with the establishment of trading companies, such as the Hudson's Bay Company, the Levant Company, and the East India Company, that would eventually control almost half of all world trade. To manage these vast enterprises, managers introduced revolutionary systems of corporate governance so that the company would be "one body corporate and politick" and would have "a legal identity and form of corporate immortality that allowed it to transcend the deaths of individual founders and shareholders."[16] The joint stock share company—ultimately governed by shareholders—was born.

The dot-com bubble of the 1990s showed what can happen when young companies lack the necessary controls to grow sustainably or report transparently. But it had plenty of notorious precedents, such as the Dutch Tulip Bubble of the 1630s and the South Sea Bubble of the 1720s, which left a trail of financial destruction. Financial markets lacked the controls necessary to curb the initial frenzy and subsequent inevitable burst. Companies were governed—but did not necessarily display good governance. Their business performance ruptured as a result.

In the U.K., the Cadbury Report of 1992 brought the issue of good corporate governance into the forefront and gave recommendations on how to raise standards around remuneration for board members and the role of nonexecutive directors in guiding decision-making. The heated response even caught its author Adrian Cadbury by surprise. "The Committee has become the focus of far more attention than I ever envisaged when I accepted the invitation to become its chairman," he wrote in the report's introduction.[17] He continued:

> The harsh economic climate is partly responsible, since it has exposed company reports and accounts to unusually close scrutiny.
>
> It is, however, the continuing concern about standards of financial reporting and accountability . . . and the controversy over directors' pay, which has kept corporate governance in the public eye. Unexpected though this attention may have been, it reflects a climate of opinion which accepts that changes are needed, and it presents an opportunity to raise standards of which we should take full advantage.

A rapid groundswell of public opinion can also necessitate change. Those who are first to react—or who are already ahead of the curve—emerge stronger from the crisis. In the U.S., following seismic corporate and accounting scandals that caused the demise of Enron and WorldCom, elected politicians introduced wide-ranging auditing and financial reporting regulations for U.S.-listed public companies in the form of the Sarbanes-Oxley Act of 2002. The reforms were designed to ensure that shareholders, employees, and the public were protected in the future from similar accounting errors and fraudulent financial practices.

The financial crisis of 2008 ensured that governance remained on the public's radar, as companies that were presumed too big to fail, like Lehman Brothers, filed for bankruptcy, while others applied for government bailouts. Some of the world's biggest companies have since needed to weather major storms caused by inadequate corporate controls, leading to stock price sell-offs, new CEOs, name changes, and high-profile apologies.

This time, the legislative response came in the form of the Dodd-Frank Act (2010). However, good governance cannot just be ex post facto compliance with new external regulations. It also must comprise internally set standards and company values.

Governance Is the Back End of ESM

Good environmental and social action by companies can often be traced back to having the right governance already in place. The story of chlorofluorocarbons (CFCs), now understood to cause damage to the ozone layer, provides a telling example.

In the 1920s, the U.S. mechanical and chemical engineer Thomas Midgley, Jr., pioneered what turned out to be one of the most harmful innovations of the 20th century: CFCs, branded as Freon, for use in refrigerators and later in aerosols. Midgley won prestigious awards for his breakthroughs before the fallout of his work became apparent.

"Midgley had a more adverse impact on the atmosphere than any other single organism in Earth's history," said the environmental historian J.R. McNeill, while the author Bill Bryson wrote that Midgley possessed "an instinct for the regrettable that was almost uncanny."

By the 1970s, emerging evidence that CFCs were damaging the ozone layer raised public awareness about global warming and skin cancer. In 1975, SC Johnson, the producer of household cleaning supplies, chose to ban CFCs from its aerosol products worldwide, becoming "one of the first companies to take a major, public stand against an ingredient that was harming the environment."[18]

The decision was met with disbelief and even anger at the time. Scientific evidence was then inconclusive. CEO Sam Johnson was advised by one of his own executives to wait for proof, rather than "acting on emotion."

Johnson replied to the criticism: "Our own company scientists confirm that, as a scientific hypothesis, [the idea that fluorocarbon propellants in some aerosol containers might be causing ozone depletion] may be possible." That was all he needed to hear.

There was short-term pain, as the company withdrew from several countries that offered no alternatives to CFCs. SC Johnson walked away from the U.K. antiperspirant market, where it was currently leader.

But SC Johnson didn't go quietly. A 1975 promotion by the company in the *New York Times* explained: "We are taking this action in the interest of customers and the public in general during a period of uncertainty and scientific inquiry. We plan to change the labels of our containers to carry the following statement: Use with confidence. Contains no freon or other fluorocarbons claimed to harm the ozone layer."

Three years later, CFCs were banned in the U.S. The 1987 Montreal Protocol was eventually signed to phase out the production and consumption of ozone-depleting substances.

"When my father decided to take CFCs out of our products, he did because it was the right thing to do at the right time," said Fisk Johnson, now chairman and CEO of SC Johnson. Twenty years later, the science that convinced Johnson to act was honored by the 1995 Nobel Prize.

The decision proved to make sound business sense, too. Company scientists were able to pivot aerosol production to propane and isobutane, which were cheaper and more environmentally friendly than CFCs. By the time competitors caught up, SC Johnson was leading the way in CFC-free products and saving money in the process.

"This win for the environment and for business proved the benefit of our commitment to treating economic and environmental concerns as interdependent," says the company today:

> We continue to lead the way in responsible raw material choices. We make sure people know what's inside our products . . . and, when needed, we make the right environmental choices, even if they aren't profitable. Most of all, we continue to let science steer our way. We won't always make the right decisions, but we will always try to act based on the best research available—and when we think it's needed, we'll act even if others don't.

From a sustainability perspective, this sounds like a story that is solely rooted in the Planet factor. However, there is also a strong People element, given the impact of ozone protection on people across the globe, along with the attractiveness of working for a company with strong values. But it's really governance that provides the cornerstone, as sustainability was led from the top of the business, creating a strong culture and clear purpose. SC Johnson was ready to go beyond compliance with existing laws and follow the values at the heart of the company.

In that respect, governance provides the engine for being a Force for Good. Sustainability is helping compliance evolve from the "department of No" into one that helps companies progress over the longer term.

Time to Operationalize

There's a feeling among some business leaders that sustainability is still a luxury. They see it as an either / or choice: either operate sustainably or make money. With shareholders to serve, the sustainability path is simply not feasible, in their eyes.

I would challenge this assumption on two levels. First, as we've discussed throughout this book, those companies that understand the business incentives of innovating as a Force for Good will prevail over those that don't. Sustainability is the Force for Good of our time. Secondly, by operationalizing sustainability, businesses enjoy the benefits of greater capital, better and more productive people, better products and processes, and greater regulatory compliance.

I believe that proactive companies look to operationalize sustainability. That approach allows them to find the right data for making smart, informed decisions.

By combining People and Planet considerations beneath the banner of sustainability, companies can bring together different strands of "doing good" that they may have been doing for many years into a single reporting system. They gain an infrastructure for good. By bringing them together, Planet and People will energize Performance.

Firms are embracing sustainability as a means of highlighting their investments and returns in this endeavor. The key is to make sustainability reports visual and digestible, which can prove a challenge for businesses without external help. Ideally, a standardized format would exist to benchmark sustainability against competitors and the industry in general.

Beyond industrial companies, nearly all sectors are measuring their impact and making changes. The C-suite is getting involved as never before, partly in response to growing public interest in sustainability and partly because they recognize the financial benefits of greater efficiency.

Companies are taking proactive steps to relieve the pressures on their business, lessen future liabilities, and make meaningful decisions. It can take time and significant investment to fully operationalize sustainability, but those who do will see the benefits in terms of people, processes and access to capital.

Why is this the time for enterprise sustainability management? As we've seen with other Forces for Good in history, culture generates change—and then businesses respond with innovation. The focus on sustainability reflects where we are in the world today.

As we'll see in the next chapter, companies are at different stages of their journey toward sustainability. Discover how far your own business has traveled—and how to pick up the pace, if needed.

Key Takeaways

- Several events have strengthened resolve around sustainability as the key to doing good in 21st-century business.
- One size doesn't fit all. Businesses need to determine which sustainability issues are material to them.
- Companies can no longer hide in ivory towers; instead, they must embrace the benefits of a diverse and contented workforce that shares their values, the People factor in enterprise sustainability management.
- Planet is a driver of product innovation, and the energy transition demanded by climate change presents a major opportunity for business.

- Governance underpins sustainability. It is the bedrock of a sustainable business.
- All companies can benefit from operationalizing their material factors as part of ESM, leading to enhanced Performance—the final element in the 3Ps.

6

THE SUSTAINABILITY
MATURITY SCALE

- Self-diagnose your company's current sustainability maturity
- Are you Getting Started, Getting By, or Getting Ahead?
- How data, expertise, and technology move a company along the scale
- The benefits of using materiality assessments to survey stakeholders
- Why visionary CEOs are the instruments of sustainability improvement

"Knowing others is intelligence, knowing yourself is true wisdom.
Mastering others is strength, mastering yourself is true power."
—Lao Tzu, Chinese philosopher, ca. 550 BCE

W hen measuring the progress made by companies on their journey toward sustainability, we find that they generally fall into one of three categories. Some are just Getting Started on their sustainability journey. Others—the vast majority of companies—are simply Getting By and staying compliant. Finally, there are those forward-looking businesses that are Getting Ahead by harnessing sustainability as a Force for Good.

In this chapter, I'll look at each category in turn and discuss ways of moving your company along the sustainability maturity scale to become a leader that uses "doing good" as an engine for long-term success. Again, I'll highlight the importance of the Formula for Good: technology, expertise, and data. The encouraging news is that Getting Ahead is easier than you might expect.

Companies That Are Getting Started

At Sphera, our consultants often work with companies that are only just embarking on their sustainability initiatives. These companies may well comply with the core regulations that affect their day-to-day business, but they are looking for a more strategic approach to sustainability.

What kickstarts their journey? It could be a request from an investor or a customer demanding that they complete an emissions report or demonstrate their sustainability credentials . . . a public relations crisis driven by negative press coverage . . . or simply a CEO who finds inspiration in an article or at a conference.

The first step is to assess the company's current maturity. On the next page are snippets from a much longer sustainability maturity checklist with sample questions. In each case, the respondent would answer Yes, No, or Not Applicable.

Does your organization . . . ?	
People	1. Promote hiring from local communities?
	2. Have systems in place for prohibition of child labor?
Regulatory Compliance	3. Have systems in place to identify noncompliance concerning breaches of customer privacy and loss of customer data, along with procedures for prevention?
Planet	4. Address material conservation and recycling with specified annual targets for use of recycled materials?
	5. Conduct product life cycle assessments (LCAs)?
Energy	6. Keep a record of energy consumption?
	7. Have a clearly defined objective and commitment to systematically reduce energy consumption?
Sustainability Practices	8. Have committees in place to address the key concerns pertaining to sustainability?
	9. Identify risks and opportunities pertaining to sustainability issues?
Supply Chain	10. Map the negative environmental impacts in the supply chain and take suitable remedial actions?

Questions in the checklist also cover biodiversity, climate change and air emissions, and water, waste, and environmental regulatory compliance, among other topics.

The above examples are indicative of the approach of the entire questionnaire. Once a company gains a better understanding of its current position based on answering the full questionnaire, it can start the process of gathering data and setting a baseline to track improvement against performance indicators. This involves answering questions like, "What are the priorities?" and "Which sustainability topics are material?" We'll look at the entire process later in this chapter.

Companies That Are Getting By

This category contains by far the greatest number of companies. Those that are Getting By with respect to sustainability will likely have competent systems for compliance but lack the internal direction and drive to use sustainability as a Force for Good. Sustainability is probably viewed as a requirement rather than an opportunity. The Getting By company asks, "What is compulsory for us? What are the regulations that we need to meet at a minimum?"

That's not to say that these companies don't take compliance extremely seriously. They usually do—and they may have done so long before the modern definition of sustainability took hold. Many of today's industrial companies started their journey back in the 1960s, '70s, and '80s. A lot of progress was made regarding environmental, health, and safety issues during that time. Labor laws were introduced that weren't easy for companies to comply with, and a huge amount of good work was done to reach the necessary standards.

But change happened in a fragmented and isolated manner. By and large, it was driven by government in a reactive sense, not by companies in a proactive sense. Sustainability was compliance-oriented, rather than viewed as an opportunity for business change.

Some companies are *barely* Getting By. They will comply with some standards but fail to comply with others. Often, their sustainability commitments are stagnating. They made some progress five or 10 years ago, but today they are opting to bask in past glories rather than take the next step forward. They eke out a one percent improvement here, or they sign a new international agreement there. They may even have pledged to reach net zero emissions by 2050, but they have no concrete action plan to achieve that target.

The more sophisticated Getting By companies will meet sustainability regulations in an organized and systematic manner, in line with industry standards. They will have proper processes, assessments, evaluations, deployment of resources, and reviews in place, along with a basic Plan-Do-Check-Act model.

They calculate baselines and track performance data across key metrics, with budgets allocated to achieve those targets. They enjoy cost savings across energy or raw materials—resulting in commendable environmental dividends—which are duly communicated to markets that care. There may be customer engagement too, but that process is linked to operations and product development. The primary focus remains tactical, rather than a genuinely strategic approach that incorporates various stakeholders. Their sustainability obligations are static and not connected to an overall strategy. They are unlikely to be operationalized for the wider good of the business.

Companies That Are Getting Ahead

For true leaders, sustainability *is* the business value case. Companies that are Getting Ahead treat sustainability as an opportunity, not an obligation. The CEO has made it a priority to understand company stakeholders' expectations and aspirations. The systems for Planet and People have become integrated and then embedded in the business to raise business Performance. The latest progress is aggressively communicated and effectively documented with advanced software systems in place, backed by accurate measurement and extensive benchmarking within the organization that results in best practices.

The Getting Ahead company gains deep knowledge of communities across the value chain as part of its desire to minimize risks and capitalize on opportunities. It engages with stakeholders to seek out trends while documenting and communicating performance to gain competitive advantage.

For the Getting Ahead company, sustainability has become a core part of business strategy and brand, with robust accountability mechanisms across all core business functions. Sustainability drives key processes, leading to partnerships with key stakeholders.

In these forward-thinking firms, it's not only the sustainability manager who talks about sustainability but also the leaders in HR, marketing, sales, and procurement. Each function wants to show how and what they contribute to the sustainability effort. These companies are communicating to their stakeholders about how their businesses can mitigate risks while enhancing their brand reputations.

These leaders build sustainability into their everyday business, taking care to remold their organizations, systems, and processes, in addition to their culture, vision, purpose, and values. Sustainability reaches beyond the factory gate or retail store to include recycling, reuse, and circularity, bringing suppliers on the journey and even working with industry peers to reset best practices.

They are best in class in terms of data disclosure and governance. They have turned good ideas into commercial products and services. Getting By companies would not run the risk of investing in initiatives such as hydrogen power or carbon sequestration, but Getting Ahead leaders seize these opportunities. They are the first to remove single-use plastics or other harmful materials across their entire systems. They are committed to future-proofing their businesses.

Their watchwords are accountability, branding, and collaboration. They want to be 10 or 15 years ahead of their peers. Their R&D teams have already planned or piloted ideas, ready to deploy them across the whole organization when the time is right.

They place a sustainability lens in front of every decision and activity, from design to development to communication.

Bringing Value to Society

The chemicals sector might not be the first place one would look for a leader in sustainability. However, BASF offers a benchmark for sustainability across all sectors, with its environmental impact assessments for the thousands of raw materials that are present in almost every product we use in everyday life. With a 150-year heritage, the company's portfolio ranges from chemicals, plastics, performance products, and crop protection products to oil and gas.

In 2013, BASF set out to raise the bar on reporting its sustainability credentials through its Value to Society initiative. The company recognized that there was no "standardized approach to determine the value contribution of a company." Worldwide, there was a lack of "consistent standards for measuring the overall impact of companies that covered economic, environmental and social aspects of business activities along the value chain." Most important, the leadership saw that taking a stakeholder approach would prove beneficial for shareholders in the longer term.

So as a way of getting ahead in terms of sustainability, BASF set out to find its own solution by creating a new measurement framework called Value to Society. This initiative measures and calculates the costs of the company's performance through both its financial and nonfinancial business impacts on society, across 12 different economic, environmental, and societal categories, from net income, taxes paid, and safety to greenhouse gas emissions and water consumption. As a result, everyone connected with the business can understand the overall impact of the company in a common language.

"How can our business contribute toward creating a more viable future with enhanced quality of life? That is a question we at BASF continue to ask as we assess the challenges and opportunities in front of us," explains Christian Heller, former leader of BASF's Value to Society program:

> We needed a new way of thinking about business performance—a holistic, value-based perspective that offers a better understanding of the impact of our activities throughout not only our operations but also those of our suppliers and customers.

Ultimately, the method has to identify, quantify, value, and demonstrate our economic, social, and environmental impacts as a whole (and their connection to one another) rather than in isolation . . . Moving from the traditional shareholder value concept to a *system value* approach, we truly value the impacts and interdependencies of society and business in a comprehensive system. This system serves as our foundation for shaping the future.[1]

Sustainability Doesn't Have to Compete with Profitability

The Corporate Knights' Global 100 ranks the world's most sustainable corporations each year, providing a respected list of leaders. The chart topper in 2023 was Schnitzer Steel, since rebranded to Radius Recycling. It started life in 1906 as a scrap metal business in Portland, Oregon, so recycling and reusing is hardly a new pursuit. However, its surge up the rankings owes much to a sustainability framework that was launched around 10 years ago.[2]

"We've been recycling for about a century," said CEO Tamara Lundgren in response to receiving the accolade. "But how do we incorporate that into a framework that our employees, suppliers, investors, and communities can identify with? Weaving the bigger picture together with the specific targets into everything we do is what has allowed us to garner this honor."

In 2019, the company took the next step forward by adopting goals and metrics such as "reducing Scope 1 and 2 GHG emissions from recycling operations by 25% from 2019 levels by 2025 and reaching net-zero GHG emissions for all operations (steel manufacturing, metals recycling and auto dismantling) by 2050."

By 2022, the company's emissions were already 24 percent lower than in 2019, thanks to several measures that massively reduced Scope 1 emissions, improved energy efficiency, and used alternative fuels to power the company's arc furnaces. Performance scores on energy productivity, water, and carbon productivity were especially impressive, as were ratings for gender diversity (five out of nine board members are women) and racial diversity among executives, while the company's worker injury rate fell significantly. The company's ranking was further improved by tying its CEO's pay to sustainability targets and by offering its people paid sick leave.

In the company's long history, 2022 was its second best financially. "Our people and planet goals are clearly not coming at the expense of profit," Lundgren says. "For a company that is 116 years old and that many consider to be 'old economy' to be recognized as a leading force in sustainability is a great example of how sustainability principles can be successfully applied to an industrial company."

How to Move Forward with Data, Technology, and Expertise

For companies looking to progress along the maturity scale from Getting By to Getting Ahead, the use of data management and analysis is a wise place to start. In simple terms, companies need to first capture their data and then put it to good use.

Start by asking, "What is the goal of the data? What would good data look like?" It could be a stand-alone measure, such as calculating the company's environmental footprint in response to the need for an investor report, or it could be a greater aspiration to drive change with sustainability. Approaching management or potential customers and investors with mere assumptions won't cut it. Only hard facts, such as cost / benefit analysis, will show improvement across various scenarios.

It pays to compile a data inventory to see what's within reach and what's missing. Consider how to get more granular with data. Take a step back and ask, "What's working and what's not?" Then focus on that. As they say, the first step is often the hardest. It's better to start slowly and do things assuredly, rather than trying to tackle the whole task at once.

One easy misstep, especially when moving into the collection of Scope 3 supply chain data, is to underestimate the amount of information that's needed. Specific data can be hard to track down. Where is it in the organization? Who owns it? Of course, it's better to make estimates based on external proxy data sets than to ignore the category altogether. But authentic data is the most reliable.

Setting the baseline is vital for working out where the business currently stands. Then the focus can shift to performance and improvement, using modules that present targets and actions for, say, five years in the future.

Let's presume the target is a 50 percent reduction of emissions by 2030. Okay, that's great. We know where the company is now and where it wants to get to. Even better, what if we replace some gas-powered vehicles with electric ones over the next three years? Sounds good. But what is the end result? The total expenditure might result in just a small net reduction in emissions. Perhaps having the good news story for customers, investors, and employees will be worth the outlay. But perhaps it might be better to look at other options. This kind of analysis should continue until the optimal decision is reached.

Less mature companies will need to supplement information with estimated results. More mature companies use evidence-based targets to scientifically prove their methodologies and to track their People and Planet concerns in an integrated way. For them, the data is genuinely reflective of what the organization is doing and what they're seeing in their communities and supply chains.

If you're gathering data solely for reporting purposes, you are missing opportunities. Getting Ahead companies go beyond reporting to operationalize sustainability factors. They utilize the data with the necessary structures in place to drive change.

Companies should also ensure traceable data and a consistent data collection method—otherwise they're comparing apples and oranges. As regulations continue to get tighter and scrutiny ramps up, irregularities might be discovered, which would reflect poorly on Getting By companies.

Those who take a half-hearted approach—viewing sustainability as something they have to do, rather than something that will benefit the organization and the world they operate in—will ask someone in accounts to cobble together data from different departments and estimate the rest. If that sounds familiar, then a different strategy is needed. A fragmented approach to sustainability will not succeed in the long run.

Leaders Must Lead

In terms of people and expertise, there needs to be major buy-in from senior management, as a lack of resources will only lead to a lack of detail. Leadership is the main cause of change in sustainability, followed by stakeholder expectations and internal pressure from within the company. Even the best data and technology are flawed without the right internal knowledge and the necessary leadership to operationalize sustainability.

Once companies have stakeholder engagement and understand which sustainability topics are material for them (as I'll discuss below), they can devise a strategy and set targets. But without the blessing of management, the company won't deploy the necessary pressure to achieve those targets. Leaders must communicate clear action plans and manage the expectations and aspirations of stakeholders.

Most important, CEOs can drive change management in the organization. The sustainability manager alone cannot generate the same traction. The CEO must personally treat sustainability as a vehicle for change—and own the journey.

The forward-looking leader will make sustainability an integrated part of the business management system. Energy targets become part of the business strategy, as do water conservation, gender diversity, and community initiatives. Value chain partnerships fall under the category of sustainability. Beneath the eye of a strong leader, departments will take pride in their sustainability projects. Operations managers will receive the backing to drive through change.[3] The sustainability transformation will be energized from both inside and outside the company.

There are many CEOs in the world today who are very talented and visionary. They are the ones now driving the sustainability initiatives. They see the need to fast-track innovation.

Professor Jacqueline Peel is an expert in the field of environmental and climate change law at Melbourne Law School in Australia. She explains:

> Business leaders with very strong incentives, particularly on sustainability issues, tend to take strong action that might exceed more widespread business practices. They're answering to reputational and social license drivers, and an increasing number of activist shareholders. They may also push for higher standards voluntarily because they are operating in a multinational trading environment and don't want to be left behind.
>
> But it should be stressed that these leaders don't represent the whole landscape of corporate activity. Regulation comes in later and is aimed at the middle to bottom tier of businesses, often with the purpose of removing uncertainty about the standards of behavior. Ideally, standards adopted by the leaders will filter down, as they impose requirements along their supply chains, although it's currently unclear as to the extent that's happening.
>
> However, given the recent poor performance of governments on climate change in particular, there's a case to be made that businesses are filling that gap from a leadership perspective.

Time to Look Beyond the Spreadsheet?

In terms of software to manage sustainability, a lot of companies still run their systems on Microsoft Excel. There's nothing wrong with that—up to a point. For a small company, Excel is probably the way to go. But as the organization grows, or as regulations get tighter, Excel's limitations become more obvious. Scenario and forecast analyses are harder to conduct. And when companies commit to science-based processes, they do need some sort of platform in place to track trends over time. Eventually, Excel isn't maintainable.

More sophisticated platforms have a lot of functionality built in to allow rapid scenario and cost analysis. As sustainability starts to rival financial reporting in its complexity, a more robust system with less risk of human error, in which everything is hard-coded, becomes worth the expense.

There are numerous sustainability software applications on the market offered by consultancies, although I would argue that it pays to choose software backed by consultants with particular understanding of the sector in which their customers operate. That way, companies in need of help will gain a more granular approach to data, rather than a superficial dusting of knowledge.

Recruitment can be an accelerator if a company is able to parachute in a sustainability expert. Good sustainability consultants will also provide an additional impetus for progress. It's also worth looking for ideas within the business—and not just in areas such as product design. Employees at every level of seniority and function may have a strong desire to make a difference in regard to environmental and social causes, so it's worth inviting their opinion. Likewise, outside stakeholders, business partners, and the local community may prove to be breeding grounds for innovation.

A word of advice for those companies looking to improve their sustainability performance with a combination of data, expertise, and technology: It's important to pull all three levers down at roughly the same pace to enjoy success. If the focus is on generating data alone, without investment in technology and expertise, then the business won't make the most of its data strengths.

Companies can and should invest in putting a great system in place to fast-track maturity. But without the right data and the right expertise, they end up with a smart new engine without any fuel and no driver. The three elements of the formula need to grow in unison and work together.

What's Material—and What's Not?

The Getting Ahead companies understand what is *materially* important to their organization from a sustainability standpoint. There's the M-word again. But it's worth repeating, as a surprising number of companies don't know their material priorities. To operationalize sustainability, companies need to understand what's material, why it's material, and then find ways to achieve meaningful targets for each material factor. Each company has different risk factors. The knee-jerk reaction is to focus on greenhouse gases, but if a company isn't energy-intensive, then Planet may come in lower on the materiality list. People and Performance could be more pertinent.

Ask about the issues that will be detrimental to the business. For a services company, it could be that keeping people happy, offering training, and encouraging thought leadership are more important than investing in a new energy strategy. For a financial services company, business conduct and ethics might also be more important than the environment. By contrast, a chemicals company will likely focus its risks and opportunities on decarbonization, water reduction, and safety elements. The future of their organization can be jeopardized if there is a major accident or environmental disaster. Their name will always be associated with it.

Some topics, such as worker safety or anticorruption policies, are almost universal, while also being critical to such an extent that they demand constant supervision. There might be other important topics, like healthcare coverage or a retirement plan, that the company already has a good grip on. Yes, they're material, but at a low level of risk that doesn't need immediate attention. Then there are the critical topics that a company hasn't yet begun to tackle, which should go to the top of the to-do list.

Sustainability is often thought of as a way of reporting good things to the outside world, but it's also about avoiding and mitigating risks. This is where obtaining the views of all stakeholders can prove especially valuable.

A materiality assessment helps a company understand what all its stakeholders think about in terms of risks and opportunities. The assessment prompts the company to ask its stakeholders what they feel is important for the success of the organization, versus what the company thinks is important for the management of the organization. For example, the business leaders may consider tax and talent as their two highest priorities, while stakeholders insist on deforestation and eco-friendly packaging.

Common challenges with the materiality assessment process include incorporating and prioritizing stakeholder views, ensuring the involvement of senior management, and extending the assessment beyond the company's own internal operations.

While it can be time-consuming to conduct thorough surveys of your company's internal and external stakeholders, such studies are the most effective way to determine which sustainability issues are most important and consequential for your organization. Combined with a gap analysis and a comparison to competitors and peers, this research will help the company see what needs to be tackled most quickly. The assessment also provides an organization with deeper insights into its operating environment and encourages more effective allocation of its resources.

Thorough assessments benefit from including as wide a range of stakeholders as possible. Members of the communities in which a business operates might also be included, as their opinions will help a business understand its social impact.

Some companies may dismiss the materiality assessment as a luxury because they think that they know their priorities already, based on what society or their own instincts are telling them. But a disciplined assessment is often an eye-opening exercise, one that is worth repeating every three years to ensure that the company is focusing on the right things, rather than chasing the rainbow. They will also help shift the sustainability needle beyond mere compliance toward driving change.

Global events, crises, and trends can rapidly emerge as topics that need to be addressed in materiality assessments. The Covid-19 pandemic was a clear example, demonstrating how material issues can surface and evolve. Some crisis-driven issues may have relatively short lifespans, but they still need to be managed. A materiality assessment helps draw attention to them.

Find Truth Within the Matrix

Properly planned and executed, a sustainability materiality assessment:

- Ensures that business strategy takes significant social and environmental topics into account.
- Identifies trends that may impact the company's ability to create value in the long term.
- Identifies areas of primary interest to important stakeholders.
- Enables different functions of the business to seize opportunities that help them stay ahead of competitors.
- Prioritizes resources for sustainability issues that matter most to the business and its stakeholders.
- Highlights areas for management and monitoring that are important but not currently addressed.
- Reveals areas where the company is increasing or decreasing value for society.

A materiality matrix helps an organization understand and present the findings of its materiality assessment in a comprehensible visual form (next page). Sustainability issues are presented in two dimensions. The first dimension represents the importance of issues to the organization in terms of their expected influence on the organization's success; the second dimension represents the relevance of issues to stakeholders and the likely resulting influence on business success. By combining these two dimensions, the materiality matrix helps leaders identify the issues that are most urgent for them to address.

Senior management's view of what is crucial may differ markedly from what stakeholders think. On the matrix, the topics closer to the top are the ones that the majority of stakeholders say are material for the company. Being selective about the challenges you will tackle is important. There might be 20 or 30 issues that stakeholders deem important, which is simply unmanageable. So

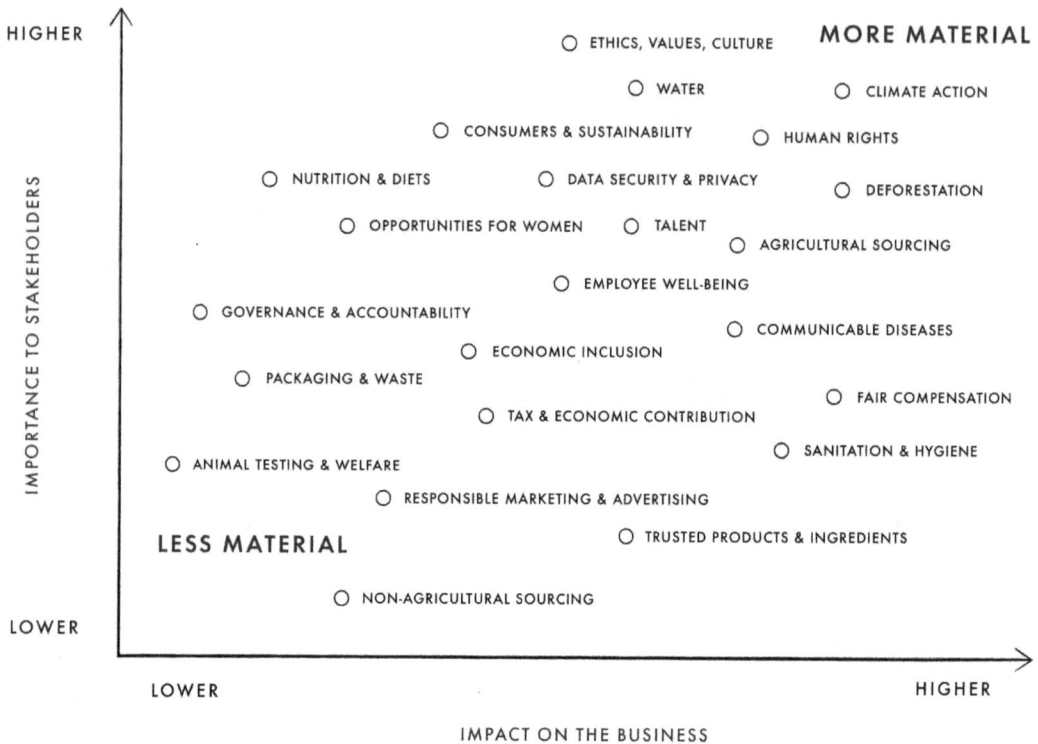

HIGHER

MORE MATERIAL

○ ETHICS, VALUES, CULTURE

○ WATER   ○ CLIMATE ACTION

○ CONSUMERS & SUSTAINABILITY   ○ HUMAN RIGHTS

○ NUTRITION & DIETS   ○ DATA SECURITY & PRIVACY   ○ DEFORESTATION

○ OPPORTUNITIES FOR WOMEN   ○ TALENT

○ AGRICULTURAL SOURCING

○ EMPLOYEE WELL-BEING

○ GOVERNANCE & ACCOUNTABILITY

○ COMMUNICABLE DISEASES

○ ECONOMIC INCLUSION

○ PACKAGING & WASTE

○ FAIR COMPENSATION

○ TAX & ECONOMIC CONTRIBUTION

○ SANITATION & HYGIENE

○ ANIMAL TESTING & WELFARE

○ RESPONSIBLE MARKETING & ADVERTISING

LESS MATERIAL   ○ TRUSTED PRODUCTS & INGREDIENTS

LOWER

○ NON-AGRICULTURAL SOURCING

IMPORTANCE TO STAKEHOLDERS

LOWER   HIGHER

IMPACT ON THE BUSINESS

THE MATERIALITY MATRIX

pick the top six or seven—no more than 10—that will make the most difference.

Peer comparison is critically important, as it allows companies in the
same sector to chart their progress on like terms. For example, many companies in the chemicals and automotive sectors have built up core capabilities in
sustainability reporting and management over recent decades. Other sectors,
less so. An outlier is the IT sector, which has a massive impact on climate
change due to the vast energy required by servers and data centers—yet this
impact tends to fly beneath the radar. The reason is that the IT sector is comparatively immature when it comes to sustainability efforts, although initiatives like the European Green Digital Coalition are making strides in bringing

businesses together.[4] Industry variations like this one should be taken into account when charting your own progress. In a benchmark assessment against sector peers, a company might appear to be a leader in sustainability, but it might not fare so well in comparison with companies from other sectors.

Sustainability Reporting—a Maturing Discipline

Standardization of reporting has proved a challenge for the sustainability movement. Companies have reported on a voluntary basis across different frameworks, making it challenging to compare like with like in a reliable manner. Over time, the gray areas will become clearer.

For European companies, the EU Commission has set the timeline for reporting under the Corporate Sustainability Reporting Directive (CSRD), which substantially increases reporting requirements and expands the number of companies that must report. Requirements go beyond carbon emissions disclosures, stretching across the full sustainability portfolio.[5] Thousands of non-EU companies with a presence in the EU also fall within the scope of the directive.

A major change in the CSRD is the need to report based on the "double materiality approach," which demands that companies disclose information related to their financial value *but also* information related to their impact on the world around them, particularly with regard to climate change and other environmental concerns. Effectively, the bottom line has been extended beyond financial implications to reveal how the firm's actions are material from a People and Planet perspective. All large and listed companies must post well-defined sustainability targets and then publish their progress every year, ensuring that sustainability becomes a central pillar of a company's long-term business strategy.[6]

The reach of the CSRD, which entered into force in January 2023, will include a company's supply chain, so the company must take responsibility for suppliers' actions rather than turn a blind eye to any unethical practices or damage to the environment by business partners. Companies subject to the

CSRD must also highlight which departments are leading sustainability efforts and offer data used in the annual report to independent auditors. As a result, the sustainability maturity of companies will be increasingly brought into the open.

People may be skeptical of sustainability rating systems, but they, too, are becoming more thorough and mature. There are still issues related to standardization, but these ratings will grow in importance, especially if they continue to harmonize data and pull whole sectors together.

Investors make decisions based on the information they can acquire. If a company doesn't provide information on its sustainability performance, then investors may turn to third-party sustainability rating systems to compensate for the missing essential data. If they're unable to make use of third-party ratings, they may assume the company puts little or no effort into its sustainability status, which may in turn lead them to leave that company out of their investment portfolios and take their business elsewhere.

No Time Like the Present

In the last few years alone, sustainability has risen up the agenda for almost every company. What was once a niche concern of the R&D department is now a priority in the boardroom, leading to greater allocation of resources and budget. Companies feel under pressure from both investors and customers, who increasingly are saying, "Hey, we want you to decarbonize. Do you have a human rights program? What happens in your supply chain?" Companies are being squeezed from both sides.

Let's be under no illusions. Tomorrow, everything in sustainability will be calculated in terms of dollars. There will be financial metrics for carbon, water, biodiversity, and even societal goodwill. This is coming soon, because it's what people and investors want.

Some experts make the case that mandatory reporting of the environmental impacts of corporate behavior will ultimately produce significant benefits for both the business community and society as a whole. Consider, for example, a 2023 study featured in the journal *Science* and authored by a team headed by Michael Greenstone, director of the University of Chicago's Institute for Climate and Sustainable Growth. The scholars calculated the economic cost of carbon emissions for over 14 thousand publicly traded companies from around the world, finding huge variations within industry sectors. This suggests that major improvements in environmental performance might be possible if laggard companies faced public pressure to match industry leaders. The authors concluded: "Revealing corporate carbon damages would start a public dialogue about the contribution of corporate activities to the climate problem, which in turn could spur policies and unleash market forces. Put plainly, it is difficult to imagine a successful approach to the climate challenge that does not have widespread mandatory disclosure as its foundation."[7]

When conditions are ripe, change happens quickly. Just think how quickly the smartphone revolution took hold across the world. A similar snowball effect will happen in the sustainability arena, especially around climate change. Many companies have set targets for net zero by 2050. That's still a long way off. But many have also set interim targets for 2030. That's not far away at all.

We can view the corporate world as a long race, with companies competing to differentiate themselves and succeed. When it comes to sustainability, today's Getting Ahead companies will come out on top. The Getting By companies will stay in the race, but not make much progress relative to their more mature competitors. Those that can't innovate will not complete the race. Many will be gobbled up by the large companies of the future.

The first step toward becoming a leader is to embed sustainability into business operations. This doesn't necessarily mean an overhaul of the way the company does business. It requires the acceptance that these issues are simply business issues. Energy, for instance, is a business issue.

Find those six or seven material issues that can drive your business forward. Make them part of quarterly reviews. Track the data. Understand which partners are needed in the supply chain to collaborate on new-generation products. Once sustainability policy becomes a business imperative, the company can gradually build the policies and governance needed for high

Performance. Heads of department will become change agents. Sustainability will become a company-specific management system. CEOs need to tackle those things that are real and then hold themselves truly accountable at a personal level, in addition to the business doing so as a whole.

Deeply embedding sustainability will allow the business to achieve excellence and meet both today's and tomorrow's requirements, while also satisfying the evolving expectations of stakeholders.

The Getting By companies will lose out to the Getting Ahead organizations that really understand the business benefits of sustainability, build out dedicated teams, and hire good people and good consultants to drive progress.

Is inactivity really worth the risk?

Key Takeaways

- Greater standardization and higher scrutiny around sustainability are coming soon.
- A materiality assessment, with results summarized in a materiality matrix, will help you identify your sustainability priorities.
- For Getting Ahead companies, sustainability *is* the business value case.
- Getting Ahead companies also operate according to a clear set of principles:
- Sustainability is integrated in the business.
- Company leaders gain a deep knowledge of the value chain.
- Leaders also place a sustainability lens in front of every decision and activity.
- Sustainability drives key processes, influencing and shaping partnerships with key stakeholders.

THE INVENTION TEST: PART 3

W e're heading into the home stretch—which means the third and final part of our quiz on sustainability innovators is waiting for you on the following pages. Let's see how much you know about six more people and organizations who've helped make our world what it is today.

Question 13

Who took wireless remote control to new levels in 1898?

A. Thomas Edison
B. Nikola Tesla

The answer is B: Nikola Tesla.

Today, the creativity of inventor Nikola Tesla tends to be overshadowed by that of his bitter rival and one-time employer Thomas Edison. Yet Tesla's alternating current (AC) electric motors ultimately prevailed over Edison's direct current (DC) in the so-called War of the Currents during the late 1880s, and Tesla's thinking still underpins our modern electrical system. Born in modern-day Croatia, Tesla at the height of his powers was a celebrity who rubbed shoulders with the likes of Mark Twain and J.P. Morgan.

A remote control is a device that uses electromagnetic waves to send operating signals to a distant device. Tesla pioneered the concept by demonstrating a radio-controlled boat in a exhibition at New York's Madison Square Garden. He hoped to attract funding from the U.S. Navy and predicted that the destructive power of remote-controlled weaponry would be so great it would effectively "abolish war." Today, however, his contraption is considered a precursor to drone technology.

Unlike the more practical Edison, Tesla was a boom-or-bust dreamer who enjoyed grandstanding his genius. His vaunting ambition proved his undoing, and he died in poverty and obscurity—although his fame lives on today in the name one of the world's most valuable technology brands.

Question 14

Which U.S. president patented an invention for buoying boats in shallow waters?

A. Abraham Lincoln
B. Theodore Roosevelt

The answer is A: Abraham Lincoln

A s a young lawyer and entrepreneur on the American frontier, Abraham Lincoln had a natural inclination toward finding out how machinery worked and then tinkering with it to improve it. In 1859, a year before launching his presidential campaign, he remarked in a speech that the patent laws "added the fuel of interest to the fire of genius, in the discovery and production of new and useful things." A decade earlier, Lincoln himself had demonstrated his personal belief in the power of the patent system. He was granted Patent No. 6469 for his design for a device to lift boats over shallow waters in rivers—although his invention was never actually built. It leaves him as the only U.S. president to date who holds a patent.

Although Theodore Roosevelt never earned a patent, he was an avid supporter of technological innovation in his own day, a generation after Lincoln. Roosevelt was the first president to use an automobile in a presidential procession (1902), the first to install a telephone in a presidential residence (1903), the first to send a transatlantic cable for diplomatic purposes (1903), and the first to ride in a submarine (1905). And although his support for laws (such as antitrust regulations) imposing controls on business was controversial in its time, Roosevelt made it clear that his purpose was not to stifle business innovation but to strengthen it by curbing its worst excesses.

Question 15

Which physician is credited with discovering penicillin—by accident?

A. Alexander Fleming
B. Edward Jenner

The answer is A: Alexander Fleming

B efore the invention of antibiotics, even a simple paper cut could prove deadly if infection set in. In 1928, the Scottish physician Alexander Fleming was experimenting on the influenza virus. He returned to his lab from a holiday to find an antibiotic mold growing on a Petri dish that he'd forgotten to clean up.

"One sometimes finds what one is not looking for," he later reflected. "I certainly didn't plan to revolutionize all medicine by discovering the world's first antibiotic, or bacteria killer. But I guess that was exactly what I did."

Fleming himself failed in his attempts to purify penicillin for medical use. A decade later, a team of Oxford University scientists turned the mold into a drug that has saved hundreds of millions of lives. However, there may yet be a sting in the tail. Antimicrobial resistance, caused by the current widespread overuse of antibiotics in medicine and farming, could lead to catastrophic loss of life in the future. As we'll see throughout this book, the impacts of technological developments are not always purely positive. It takes foresight, wisdom, and flexibility to insure that innovations are applied in ways that continue to benefit humankind.

Dr. Edward Jenner's innovation did not emerge from by accident. He created the world's first successful vaccine after spotting that milkmaids infected with cowpox were immune to smallpox. Described by the World Health Organization as "one of the most devastating diseases known to humanity," smallpox had caused millions of deaths over the previous 3,000 years. Thanks to Jenner's vaccine, the appalling condition was eradicated in 1980, making it the first infectious disease eliminated by human ingenuity.

Question 16

Who was the first person to win two Nobel Prizes?

A. Marie Curie
B. Albert Einstein

The answer is A: Marie Curie

Marie Curie, the Polish-born French physicist, won the Nobel Prize for Physics in 1903 with her husband Pierre Curie and Henri Becquerel. After Pierre's death, she went on to win the Nobel Prize in Chemistry in 1911 on her own. The first prize was awarded for the discovery of polonium and radioactivity, while the second was given for the isolation of pure radium. Curie's work has greatly affected the fight against cancer around the world.

Curie is one of only two people to have won Nobel Prizes in different categories (the other being American Linus Pauling, who won in Chemistry and Peace).

Albert Einstein won a single Nobel Prize (in Physics). He received this honor in 1922 for his explanation of the photoelectric effect, a cornerstone technology of the solar energy revolution today. Despite widespread public expectation, Einstein never received a Nobel Prize for his general theory of relativity, usually considered an even more groundbreaking scientific achievement.

Question 17

Which business had the opportunity to purchase Netflix for $50 million in 2000?

A. Amazon
B. Blockbuster

The answer is B: Blockbuster

At the turn of the millennium, entertainment startup Netflix was reeling from the dot-com crash, which decimated the prize of many tech stocks. The company's DVD-by-mail rental model was beginning to look increasingly frail. Meanwhile, Blockbuster was the top movie rental store in the U.S., with over a quarter of market share and around 87 million customers. Worldwide, the chain employed over 80,000 people across 9,000 stores and was nearing its peak value of $5 billion.

Under the circumstances, Netflix's offer to merge with Blockbuster for $50 million seemed reasonable—and in retrospect, it would have been a brilliant move for the industry giant. But Goliath rejected David's proposal.

Just 10 years later, Blockbuster had filed for bankruptcy, unable to compete with online services and the rise of video streaming. Netflix, meanwhile, had built its subscriber base with instant access and binge-watching in mind. Staying ahead of the curve, the company grew at an astronomical rate during the 2010s. It also reinvented itself as a TV and movie studio, creating a one-stop shop for content. By 2017, Netflix had more subscribers than all the cable TV companies combined.

Netflix today has a market cap of $271 billion. Meanwhile, there is just one Blockbuster store remaining, located in Bend, Oregon. Curiously, this store became the subject of a Netflix series.

Amazon offered to buy Netflix for $15 million in 1998, but Netflix's founders held their nerve. In fact, the meeting with Jeff Bezos persuaded them to focus on rental, as they realized that they would never beat Amazon in terms of online DVD retail. As of early 2025, Amazon Prime's streaming and production service is just marginally ahead of Netflix for U.S. market share.

Question 18

Which 1954 technological innovation revolutionized the viewer experience in pro basketball?

A. The shot clock
B. Live televised games

The answer is A: The shot clock

Before the introduction of the shot clock, professional basketball was boring many fans to sleep. The nadir came in the 1953 NBA season, when the Fort Wayne Pistons hogged the ball against the Minneapolis Lakers, scoring a mere 19 points to win. The league needed to change or die.

Syracuse Nationals' owner Danny Biasone saw the writing on the wall. Recognizing that change was coming, he made his players practice against the clock even before his proposed 24-second shot rule was adopted by the NBA in 1954.

Not only did the total points-per-game average immediately jump from 79 to 93 (it's now 115), but fans could enjoy increased tempo and fresh jeopardy with every play. The clock was eventually mounted on four-sided blocks above each hoop, making it easier for players and spectators alike to keep track. Thanks to the shot clock rule, the action was more compelling for home viewers when pro games were first broadcast live in 1965, and other sports have since introduced their own variations on the shot clock. Imitation is a sure sign of good innovation.

Across sports, technology has brought a wide range of improvements to make contests safer, fairer, and more exciting, from enhanced field conditions and more sophisticated practices for player diet and conditioning to strategic data analysis and updated stadium designs. The changes were often opposed at first, but most quickly became accepted facets of the game. Imagine the Olympics or Super Bowl without instant replays, or Wimbledon without graphite rackets and Hawk-Eye line calls.

7

ACTIVE STAKEHOLDER

IN YOUR ECOSYSTEM

- Adjusting to an outside-in mindset
- Turning supply chains into value chains with data
- It pays to find the overlaps between sustainability and business strategy
- How companies get the most from life cycle assessments . . .
- . . . To better understand the workings of their ecosystem

"You can't just keep doing what works one time. Everything around you
is changing. To succeed, stay out in front of change."
—Sam Walton, founder of Walmart

In the 1930s, British botanist Arthur Tansley coined the word *ecosystem* to capture a community of living things cohabiting, competing, and collaborating in the same environment of earth, vegetation, air, and water. He noted how, given limitations on space and resources, these organisms had to work with their ecosystem to exist and thrive. Those that can't co-evolve will become extinct, sooner or later. Some are more dominant species, setting the beat for others to follow, but they too are ultimately doomed if they ignore the environmental realities of their ecosystem.

Sixty years later, in his *Harvard Business Review* article "Predators and Prey: A New Ecology of Competition," the business strategist James Moore brought Tansley's idea into the boardroom.[1] He painted companies as organisms that must likewise adapt, evolve, cooperate, and interact to get ahead or even simply to survive. The more that members understand the workings of their ecosystem, the greater the potential for them to flourish. Innovation is made easier. More value is created collectively than individually. Risk is mitigated by working together.

One of the major benefits of a business ecosystem is its capacity to "lift all boats" on the tide of vital commodities such as ideas, talent, and capital. Social and environmental challenges are especially relevant, as meeting them is often too ambitious a change for a single company to achieve on its own.

So far, we've portrayed stakeholders as outside forces that have an influence on a business, like planets orbiting the sun. In this chapter, however, I'll discuss the need for businesses to be stakeholders in their wider ecosystems. This outward focus is in the best interests of the company, as ecosystems generate a value pool of ideas and opportunities. By playing an active role in their ecosystems, companies can access the good work of others and reduce the need to initiate and scale sustainability-related innovation alone.

Some ecosystems are clearly defined, such as Silicon Valley or fintech and biotech hubs. However, less formal ecosystems also combine on a sector or geographical basis. In fact, every business is part of a wider ecosystem—it's just a question of how fully it wants to participate.

When taking an ecosystem approach to sustainability, the place to start is your own purpose, which can be treated as the North Star to keep you on track. Simply put, why are you in business? For most business, over time, the *how* and the *what* may alter, but the *why* stays the same. Clearly defining a meaningful purpose has proven to be a telling factor in converting short-term growth into long-term growth. As noted by Mark Weinberger, former Global Chairman and CEO of EY, "Companies with an established sense of purpose— one that's measured in terms of social impact, such as community growth, and not a certain bottom-line figure—outperformed the S&P 500 by 10 times between 1996 and 2011."[2]

Blurring the Lines Between Sustainability and Corporate Strategy

Setting a good intention alone doesn't guarantee success. Sustainability needs to make commercial sense. Companies need to ask, "What are the sustainable actions that will create value? Where's our business model in relation to societal and environmental issues? Which of these issues could be a cost play or a revenue play?"

For example, take a business with high energy costs. Energy efficiency has commercial benefits. It also happens to be better for the planet by generating fewer greenhouse gas emissions. So the business may cooperate with competitors—in sustainable *coopetition*—to develop a renewable solution that reduces energy costs for all. In the process, they also generate a good news story that is beyond their individual means.

As an example of coopetition, the food companies Danone and Nestlé Waters teamed up in 2017 with the U.S. company Origin Materials to design plastic bottles made from 100 percent bio-based materials (cellulose fibers) by 2025. PepsiCo joined a year later.[3] By seeing the bigger picture, all three brands planned to benefit—while aligning with the ecological concerns of many of their customers.

For an employee-driven business, environmental factors such as energy consumption may be less critical cost metrics than, say, staff turnover. The need to train new people inevitably affects efficiency and productivity, as newcomers take time to learn procedures. High turnover can prove frustrating for customers, too, who experience more errors, delays, and the same questions that the previous trainee asked last month. Customer satisfaction suffers. Research shows that retail stores with the best rates of staff loyalty also score highest in customer satisfaction.[4] Those companies that treat their people well retain them, which is good for the customer and good for investors, too. Thus, thinking from the perspective of all the stakeholders in your ecosystem—including employees, investors, and customers—is good for business.

There is often a natural affinity between purpose and sustainability, as an effective purpose is generally linked to social issues. By solving a societal problem, companies create demand for their products. Profit and sustainability are therefore linked, too. Sustainability issues are business issues. Through this lens, sustainability is a business strategy that aims to maximize investor returns, often over the longer term.

Another way of looking at it is to say that sustainability is innovation for the future. If you carry on making the same things in the same way for the next 10 years, then the chances are that your market share, quality of staff, and investment will erode over time. In today's changing world, that erosion may happen far quicker than before. By contrast, viewing sustainability as just another form of business innovation—a bet on the future—transforms sustainability from a cost into an opportunity for advancing the company strategy.

Taking an ecosystemic approach isn't always the easy choice. Thinking from the outside in can feel culturally awkward. Companies need to painstakingly analyze their influence on the world. They should regularly ask themselves, "What are the things the customer cares about most? Which events might cause our people or investors to jump ship? How does our current business trajectory play out? Will it end well? And if not, what needs to change so that it will?"

For an example, look at the way Amazon is now employing large teams of sustainability professionals to improve its packaging and delivery experience. Customers are irritated by the amount of excess cardboard they need to recycle and by the doorbell buzzing five times a day. If enough of those customers think, "This has to stop. I'm going to shop in-store once a week instead of ordering everything online. It's more convenient, and I'll feel better about it," then Amazon is going to lose out. To avoid such a backlash, the company is taking proactive steps. Less packaging and fewer deliveries are also sustainability wins. There is a clear overlap between the business and sustainability. In the words of Kara Hurst, chief sustainability officer at Amazon, "Sustainability is now a critical consideration in decision-making from the boardroom on down, and customers demand that we understand these issues as a core part of how we run our businesses and operate in our communities."[5]

I think that companies should obsess about finding those commonalities between business and sustainability. Why is our business a Force for Good in the world? Imagine a precocious child who keeps asking *why*: "Yeah, but why? Why?" Sometimes you need to ask four or five *whys* before you get to the heart of the challenge.

It's up to the chief sustainability officer to guide those conversations internally—or to lean on outside consultants to bring their insight into sustainability strategies. In time, the lines between the head of sustainability and the head of corporate strategy could become increasingly blurred. As employee and customer power continues to grow, which it will, sustainability becomes less about doing good and more about just doing business. Meanwhile, the ties between the company and the wider ecosystem will grow stronger.

Thinking Outside-in with Renewed Purpose

Walmart, the world's largest corporate employer (2.1 million employees) and largest company by revenue ($640 billion), was founded in 1962 by Sam Walton with a simple purpose: to "save people money to help them live better."[6] As he transformed his family's five-and-dime in Bentonville, Arkansas, into a global discount chain, Walton's aim was to help people save money so they could live better.

Over the following decades, the company grew and fulfilled its purpose to the benefit of customers, who had more money in their pockets at the end of the week. Sam Walton is regularly honored as one of the most influential business leaders of all time.

But size brings its own problems. Walmart became increasingly viewed as a cause of societal and environmental damage. Walmart and other big-box stores provided a large target for global concerns about waste and the low rewards to labor. Walton was once admired for his ability to squeeze suppliers on behalf of customers, but over time his company was disparaged by critics as the "Bully of Bentonville."

By the mid 2000s, Walmart's public perception was starting to hurt its profits, as smaller discount stores ate into its market share. A 2006 marketing report by Walmart's then-advertising agency described the public's view of the chain as that of a "bad corporate citizen who doesn't treat employees well and isn't acting as a good citizen of the planet."[7]

Walmart didn't need a sustainability rating to recognize where it was falling short. The company took action—and it did so by returning to its core purpose. In what has since been described as a green revolution, Walmart has found new ways over the last 15 years to save people money and help them live better.[8]

Some of the innovations have proved industry-changing, not least because of Walmart's reach and influence. For example, the retailer formed the Sustainability Consortium, a global nonprofit that aims to transform the consumer goods industry to deliver more sustainable consumer products by bringing together manufacturers, retailers, suppliers, service providers, NGOs, civil society organizations, governmental agencies, and academics. The 100+ members include organizations as varied as Amazon and ExxonMobil Chemical, the World Wildlife Fund, and the Sustainable Apparel Coalition.[9]

In 2017, Project Gigaton was designed by Walmart to remove a billion metric tons of greenhouse gas emissions from its supply chain by 2030. By 2022, the company was over halfway there.)[10] More than 4,500 suppliers, comprising more than 70 percent of the company's U.S. net sales, joined the project—one of the largest private-sector emissions initiatives ever. Walmart's sustainability goals include zero emissions across global operations by the year 2040 (without carbon offsets), as well as 100 percent recyclable, reusable, or industrially compostable private-brand packaging by 2025. The list of other targets is eye-catching: 100 percent renewable energy by 2035, zero vehicle emissions by 2040, and low-impact refrigerants for cooling and electrified equipment for heating by 2040. Walmart has pledged to protect, manage, or restore at least 50 million acres of land (the size of Nebraska) and a million square miles of ocean (the size of Argentina) by 2030.[11]

Walmart's ambition shows how large, dominant organizations in an ecosystem can make an outsized impact on good practices. This is not simple philanthropy, or even a defensive push for popularity. Global warming is a direct threat to the commodities that Walmart needs to source from around the world. It means higher energy prices, which means higher prices at checkout. Bad for the planet is bad for business. Meanwhile, less packaging, water, or transport saves millions of dollars, which can be passed back to the customer.

While marking the company's 60th anniversary in 2022, CEO Doug McMillon summed up the philosophy this way:

> As we engage and serve our stakeholders, we grow and strengthen our business—and vice versa. If the past couple years [of the pandemic] have reinforced anything, it's the need for this kind of shared value approach. At Walmart, this isn't something sprinkled throughout our business or added on as an afterthought—we strive to use this approach in all aspects of our business and help guide us. Throughout our journey, we have set goals, measured progress and reported on that progress. And we continually push ourselves, which is why we elevated our aspiration by making a commitment to become a regenerative company—one that puts people and nature at the center of our business practices.[12]

McMillon's use of the word *regenerative* is important. Where a sustainable firm takes steps to reduce its ecological "footprint," a regenerative company seeks to grow its socio-ecological "handprint" by helping people and the planet to recover, according to Harvard Professor of Public Health, Gregory Norris.[13] In doing so, regenerative businesses will demonstrate greater financial performance and social impact than even those companies with a clear sustainability focus.[14] If Walmart fulfills its ambition to become a regenerative company, then the benefits will be felt far across its massive ecosystem.

Regenerative companies are like forest ecosystems. We now know that trees pass nutrients to other entities through an underground fungi network (inevitably dubbed the *wood-wide web*).[15] They give back more than they take out. This seemingly altruistic behavior sits at odds with the traditional view of a selfish, extractive big business—but if it works for trees, why not for companies?

Walmart isn't there yet. The global chain still has its critics. But it's also true that many once vocal opponents are now active members of the Walmart ecosystem, as they see the scale of impact that's possible for such a vast organization.[16]

Sparks Fly When Sustainability and Business Sense Collide

Companies need to stay close to their core source of business value. This mission goes far beyond making a catchy marketing statement. It's the philosophy you subscribe to. If your employees are miserable, you will have high turnover costs. If you serve cold gumbo or the lines at the store are long on the weekends, then customers are going to walk away—and grumble to their friends and online community. So it's a case of being aware of everything that's going on and then linking it to business value. That's a much better approach to playing a role in society than merely making a public statement about changing the world.

Wegmans Food Markets is a family-owned supermarket chain in the U.S. with a loyal customer base. Several decades ago, Wegmans recognized a simple truth: Customers want to pay for their groceries as quickly and as smoothly as possible. So management thought hard about the best way of doing that. And they innovated. For a start, they pioneered the idea of putting the bag in front of the clerk, rather than at the side, so that the process works much more efficiently. Simple, but very effective. Then, in the 1970s, Wegmans championed the introduction of barcode scanning.[17]

Wegmans also saw the benefits of employing bright high school students in their grocery checkout lines. But how do you get those good kids to stick around after they graduate? Wegmans said, "If you come back to work for us in the holidays after your freshman year, we'll give you a college scholarship." Bingo! Now they could offer customers a world-class experience and help families in the local community. Shoppers come for the low prices and they stay for the excellent customer service.

Since 1984, Wegmans has given out $130 million in scholarships to more than 42,000 employees.[18] "This program provides growth opportunities for our people, helping them pursue higher education and career goals that positively benefit communities," explains Colleen Wegman, CEO and president of her family's business. "We have always believed that if our people are valued and supported, they will give their best to our customers and to each other. I am grateful and proud of what they are able to accomplish every day."

A Force-for-Good perspective permeates Wegmans' business, whether it's the farmers-market-style displays of fruit and vegetables to promote healthier eating, the Perishable Pickup Program that donates unsaleable produce to food banks, or the commitment to phase out single-use plastic bags and other single-use plastics as part of its drive to reduce in-store plastic packaging made from fossil fuels by 10 million pounds by 2024. Wegmans was also a leader in maintaining links with customers during the Covid-19 pandemic and then reopening safely when permitted.

An idea like the scholarship scheme doesn't happen by accident. It may seem obvious with hindsight, but it takes deep thinking to recognize that customers want an easy, efficient experience from knowledgeable staff—and that finding people like that is hard at low wages. Wegmans devised a solution that was good for business and for society, too. It's worth noting that this sustainability mentality stretches across the different departments of the company, from finance and HR to marketing and customer experience. It's not just a sustainability campaign, but a part of the corporate strategy.

Profit and Sustainability Can Walk Hand in Hand

In their bestselling book *Net Positive: How Courageous Companies Thrive by Giving More Than They Take*, the authors Paul Polman, ex-CEO of Unilever, and sustainable business guru Andrew Winston explain why businesses can lead the world to a better future—and profit through fixing the world's problems. Their premise is that a "net-positive company will improve the lives of everyone it touches, from customers and suppliers to employees and communities, greatly improving long-term shareholder returns in the process." Polman continues:

> By taking ownership of all the social and environmental impacts its business model creates, a net-positive company will provide opportunities for innovation, savings, and building a more humane, connected, and purpose-driven culture. It partners with competitors and government to drive transformative change that no single group or enterprise could deliver alone.
>
> The first step out of the safety of your own operations is to reach out, in genuine partnership, to your direct value chain. It's the start of expanding your sense of ownership—net-positive companies don't outsource their life cycle responsibilities. This is the biggest immediate unlock in value, given the much larger total footprint in the value chain. There are often enormous savings or higher revenues if you work better, in trust and cooperation, with suppliers and customers.

Polman was ready to walk the walk. He and his team launched the ten-year Unilever Sustainable Living Plan that set several ambitious targets for 2020: helping more than a billion people take action to improve their health and well-being; halving the environmental footprint for the making and use of their products as they grew their business; and enhancing the livelihoods of millions of people. Unilever achieved most of its targets while raising revenue by a third.[19] Across Polman's tenure, shareholders saw returns increase by nearly 300 percent.

Statistical analysis software company SAS is another firm that is held up by the business community as a benchmark for how to lead successfully over the long term. The company was founded in 1976 by CEO Jim Goodnight, who has overseen sustainable growth for the best part of 50 years, leaving plenty of competitors on the side of the road.

What's kept SAS on the high road? Partly, it's the dedication to research and development, with a quarter of the company's annual expenses invested in determining what the customer will need next. This forward focus has led to several technological pivots without which the company would have faltered. Through strong governance, the company has taken care to grow within its means, gaining the resilience required to endure market slowdowns.

Goodnight also sticks to a golden rule: "If you treat employees like they make a difference, they will make a difference." This level of respect also results in very low turnover and strong showings on good employer leagues. Besides, it's "the right thing to do," he says.[20] Goodnight explains:

> We have a fairly young workforce. And we want to encourage them to be out there exercising and staying healthy because, in the long run, we have to pay their healthcare costs. So, it's better to try to keep them healthy to start with. And we have our own medical facilities on campus. . . . So, any problem you have, you just go over to healthcare and usually within five or ten minutes, you see a doctor and you're on your way.[21]

Support services provided to employees by SAS range from prenatal guidance to marriage counseling, locating care homes for aging parents to hairdressing. Goodnight continues:

> I think the job of CEO is to help alleviate as much stress as possible on the people that work there. I mean, we want some good stress. I mean, you know you got a deadline that's coming up. But we don't

need the stress like *oh, I've got to leave and go to the dry cleaners.* So, we try to do as much of that kind of stuff as we can right here, right on campus. I often say that 95 percent of my assets drive out the front gate every night and it's my job to make sure they come back the next day.

SAS shows the clear crossover between social good and business strategy in the way it trains up the next wave of data analysts. "For decades, SAS has helped students build skills to succeed in the classroom and beyond," says Goodnight. "Our data literacy initiatives challenge students to explore solutions to the world's problems, tapping into their passion and energy while instilling in them a deeper understanding of data." Students who catch the data bug have since gone on to work for the company itself. "With data and analytics," Goodnight says, "we can see the world through a powerful lens. Whether in service to our community, interpreting the news or championing the U.N.'s Global Goals, when we understand data, an entirely new world opens to us."

Life Cycle Assessments: The Bridge into Your Ecosystem

In 1987, the United Nations Brundtland Commission defined sustainability as "meeting the needs of the present without compromising the ability of future generations to meet their own needs."[22] To achieve this goal, companies need to understand their impact on the world in which they operate—and draw in knowledge from their ecosystems.

This is hard without the right tools to gauge current performance and measure progress. That's where life cycle assessments (LCAs) can make the difference. LCAs allow companies to compare and improve products and policies. They send feelers up and down supply chains, searching out valuable information. Done right—as I'll explain below—LCAs provide much more than just a safety net against regulatory or reputational risks—although these are both important. A rigorous LCA will also surface new value streams across your ecosystem—for customers, suppliers, employees, and the wider environmental landscape. The assessment can help the business become a Force for Good. What's not to like?

LCAs are commonly used as a data-generating tool for external reporting. Again, this is helpful, but it falls short of the full potential of an LCA as a powerful engine for internal decision-making. As with sustainability in general, the full benefits of an LCA are realized only when a business operationalizes the findings. That's when material change happens.

An essential benefit of an LCA is that it encourages the business to look for value beyond the perimeter fence. Traditionally, the information spotlight at most companies was shined on internal concerns alone, such as manufacturing, logistics, and labor. Materials arrived at one gate; goods and waste left through another. Interaction with the outside world took the form of marketing, PR, and recruitment. Progress was measured solely in dollars and cents. What you didn't know couldn't hurt you.

An LCA is designed to add texture to the traditional one-dimensional view into supply chains. Relevant data can turn supply chains into a source of opportunities. Of course, there is a risk that investigating the provenance of goods, carbon emissions, diversity and inclusion, waste management, and the like will reveal unpalatable truths. Reaching out requires an information-sharing culture that can make some CEOs feel queasy. Finance directors may balk at the possible time and cost implications, especially as change is often needed to realize benefits from the LCA's findings.

Yet the potential upsides outweigh the shortfalls in terms of cost efficiencies and improved reputation within the stakeholder ecosystem. Any argument that "ignorance is bliss" will lose its appeal as sustainability considerations continue to grow in importance. LCAs spark innovation, which is a good thing.

The LCA is more than a footprinting measurement, which tends to limit the focus on carbon emissions and water use. Yet LCA methodologies can be used to calculate product carbon footprints that quantify greenhouse gas emissions for each stage of a product's life cycle. Proactive companies use these assessments to identify carbon hotspots in their products' value chains, gaining data-driven insights for communicating with suppliers, customers, and other relevant stakeholders. Depending on the company's sustainability goals, these can be used to achieve emissions reductions and advance a decarbonization strategy toward net zero.

LCA Data Draws Its Power from Industry Knowledge

Guided by a framework drawn from the globally accepted ISO 14040 and 14044 standards, an LCA takes you beyond carbon emissions and provides your company with a big picture of your product's environmental performance. To do this, specific understanding of the industry in which you operate is crucial. Sphera specializes in the field of sustainability data software, and I've seen firsthand how LCA consultants with the necessary depth of industry understanding can aid a business in focusing on the sustainability indicators where improvements are most needed. They can ensure that environmental burdens are not just redirected to have impacts in other, less obvious areas. The range of options is growing, such as calculating water footprints of products or conducting social LCAs, life cycle costing, or comparative LCA studies to improve product sustainability with optimal efficiency.

As the name suggests, the LCA takes you through each step of the product or service life cycle. How are raw materials extracted? How much energy and water do you and your suppliers use, and where does it come from? How are these resources used in design, manufacture, packaging, and transportation? Who is doing the work? What are their working conditions like? What's the downstream impact of your products and services? What happens to the waste and pollution you create in production and at the end of product life? What volume of materials and resources are returned to the life cycle?

Taking a more active role in the wider ecosystem with a comprehensive LCA can result in sustainable improvements that will be reflected in the bottom line. Cost savings is at the top of the list of benefits. The LCA performs an audit of the resources your company uses, many of which may be repeat orders dating back over many years. Shining a light on supply chains may prove mutually advantageous. By contrast, a limited mindset risks causing stagnation for all parties along the chain. There may also be opportunities to make savings through material recycling and reuse and by diverting assets that were previously assumed to be garbage away from landfills. Data and software may reveal new avenues for efficiency in procurement, operations, distribution, and disposal. For example, a business might run a life cycle cost analysis across several potential product designs, identifying the one that is least expensive from both a carbon and financial perspective.

The cost savings can be attractive in the present—but also in the future. With greater knowledge regarding resource use and continuous efforts to reduce consumption and waste, companies can make themselves more resilient to, say, sudden spikes in the price of energy or water or pressures from new regulatory demands.

In the Battle for Hearts and Minds, Labels Matter

Brand identity enhancement represents another advantage of a detailed LCA. Just think how many foods and beverages, household goods, and cosmetics broadcast their sustainability credentials on the packaging. B2B businesses are also proud to publish their improving track record on their websites and annual reports. For marketing teams, this quantitative and qualitative information is extremely valuable, as it offers a route to differentiation and positive coverage.

For example, a 2021 survey from Fairtrade International and GlobeScan revealed more than half of the 15,000 respondents from 15 countries said they had changed purchasing choices over the previous year in order to have an impact on economic, social, environmental, or political issues. Meanwhile, 64 percent of those who recognized the Fairtrade label said they were willing to pay more to ensure producers earned a fair price.[23]

Having clear data (ideally in real time) is also important for winning hearts and minds among consumers and the local community if the need arises. Making claims and setting goals, both helpful practices for creating a dialogue with stakeholders, can ring hollow without the robust information to back them up.

Increasingly, information gathering and reporting are no longer a choice but a necessity. Sustainability indices, linked to the stock market, are growing in importance as a barometer for investors. For example, the Dow Jones Sustainability World Index of S&P Global covers the leading 10 percent of the largest 2,500 companies based on "long-term economic, environmental and social criteria."[24] Membership in this prestigious index sends a clear message to the outside world—as does expulsion.

From water heaters and refrigerators to the produce aisle or the meat counter, our purchasing decisions are being influenced, in part, by labels affixed to many common products. These sustainability stickers may be indicative of efficiency ratings, warranty length, nutritional benefits, country of origin guarantee—the list goes on. Whether consciously or not, we see them all the time, and they are playing an increasingly significant role in purchasing decisions.

Regulations in Europe and Japan mandate the use of sustainability labeling so that consumers can compare and contrast products. It's not hard to imagine a world that in a few years will see all products with their own sustainability stickers providing ratings based on brand, production, supply chain, and life cycle activity. Environmental sustainability is one incredibly important measure. But beyond that, social and governance topics, including worker safety and job satisfaction scores, diversity among employees, and impact on local communities, may also impact brand ratings.

Product labeling is the visible proof of a brand's journey to a more comprehensive approach to sustainability—but how the brand reaches this stage is a critical consideration. For companies to demonstrate their commitment to a more sustainable future, historical business practices must change. A strong, clear, and achievable vision for sustainability is required, along with the right software, data, and expertise to deliver products that truly improve the future for everyone.

The time for empty promises and vague marketing claims about the environmental performance of products is over. Governments, B2B and B2C customers, investors, and the public want to see accurate measurements of the environmental impacts of goods and services. LCAs help to address this need by using powerful data to crunch and then visualize inventory data. Companies are increasingly turning to sustainability software and advisors to help future-proof their products and meet the information demands of stakeholders.

Say Cheese

How exactly do LCAs work? And why do they help businesses get ahead? The answers are hidden in a block of processed cheese.

The life cycle of cheese starts with the cow, right? Well, not exactly. It really starts before the cow with animal feed. Actually, that's not right either. It starts even earlier, with soil, pesticides, fertilizer, and water. Crops aren't the same everywhere. They differ depending on the region, climate, slope, rotation machinery, irrigation, and runoff. And what about the land? Are the crops grown on arable land? Or in rainforests turned into cropping or grazing land?

There's already a ton of data in the story of cheese before we even get near the cow. But let's think about her for a moment. In addition to feed and water, she may need antibiotics. And the farmer needs electricity and gas to run the milking machines, tractors, and other equipment. We know the cow produces milk, but also lots of methane and manure. The cow may give birth to calves. And at the end of her life, she might provide meat, either for human use or for use as pet food.

But at least we now have the milk to make the cheese. Do things get simpler? Not really. For each farm that provides milk for our cheese, we need to understand how the milk is transported to the factory from the dairy. Milk tankers use different kinds of fuel and refrigeration, with varying costs and levels of emissions, depending on a range of factors.

The processing factory uses power. To fully understand the cheese, we need to know what sort of power station it comes from. And cheese uses other ingredients as well, like salt and rennet and bacteria culture. Where do they

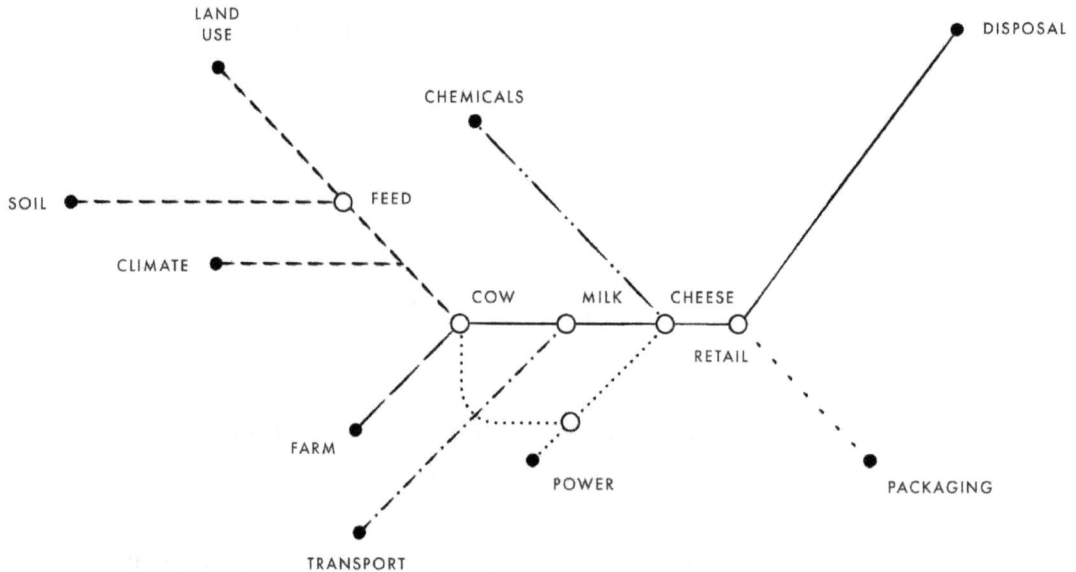

THE LIFE CYCLE OF A BLOCK OF CHEESE

come from? How were they produced? And what about waste and wastewater? Where do they go?

But finally, we have our block of cheese. Now it needs to be packaged. This packaging could be derived from oil derivatives or trees. Where do these ingredients come from? Will the cheese be eaten or thrown away? Landfilled or composted or made into biogas? The story of the cheese only ends when the cheese is truly gone.

The story of cheese, then, is surprisingly complex, as the diagram above suggests. Why does all this matter? Because we live at a time when customers, employees, shareholders, and governments want businesses to be more sustainable and reduce their production footprint. Once companies realize where the hotspots are in the product or service life cycle, they understand how to better manage the related risks, the savings this can bring, and the market opportunities it can offer. Understanding the fascinating life cycle of cheese—

or biscuits or lasagna or batteries or automobiles—allows companies to make better decisions. With access to encyclopedic data, the right software, and deep industry expertise, understanding the product life cycle results in the right decisions to use the right suppliers from the right places, to utilize the right technologies and the right materials with the right packaging and the right shelf life. With a good LCA, you can make and market the right products.

How to Make LCAs Work for Your Business

Before starting an LCA, it's worth taking some time to agree upon the boundaries of your assessment. What's in and what's out of the scope of the LCA? It goes without saying that widening the boundaries will affect both the quality of the data and the potential for good, along with the time and cost of getting it right.

For a manufacturing firm, direct impacts such as the use of resources and the amount of waste generated are staples. The bigger questions are around the indirect factors, such as the greenhouse gas emissions released when powering the factory or the community impact of mining and refining raw materials. Do you include transportation and the whole-life energy costs of your products after they have left the factory gates? Capturing and including the data from your suppliers and even customers is another step further, creating new challenges in terms of ensuring accuracy. By gathering knowledge, companies can assess and avert risk, while finding opportunities for collaboration and innovation.

Common boundaries for life cycle assessments include:

- *Gate to gate*, which focuses on the impacts caused by manufacturing
- *Cradle to gate*, which offers a wider view, starting with material extraction and ending when the final product leaves the factory
- *Cradle to grave*, which takes the whole life cycle from extraction to end of life

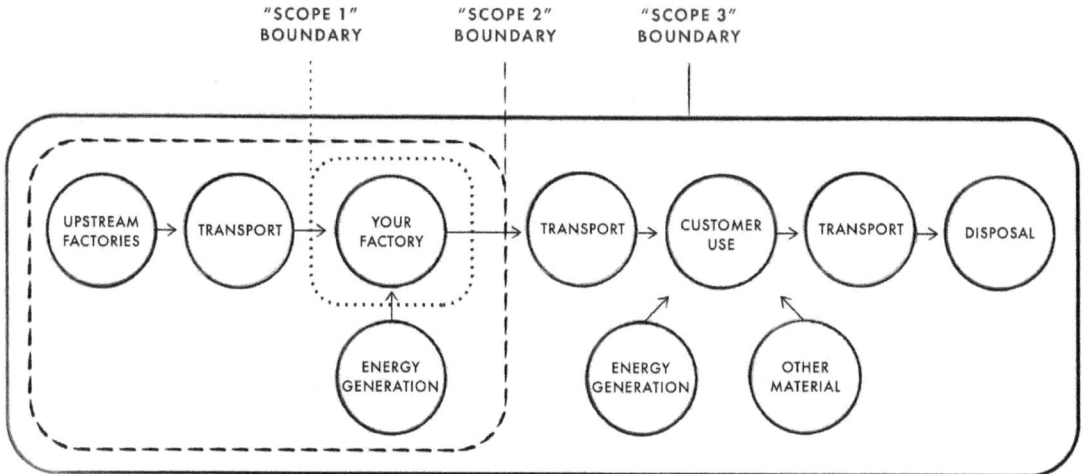

SCOPES 1, 2, AND 3 OF A LIFE CYCLE ASSESSMENT

We can see how these three boundaries play out in the figure shown above.

The first credited LCA was conducted by Coca-Cola in 1969 when the company set about analyzing the environmental impact of its drink containers. This groundbreaking approach helped to measure the cost and availability of natural resources in terms of internal costs and external perceptions. The study compared returnable glass bottles, aluminum cans, and plastic bottles, although the last-named packages weren't introduced for another decade.[25] (An independent 2020 study ran LCAs of packaging for sodas and ranked aluminum cans as the most environmentally friendly, followed by plastic and then glass.[26])

Today, LCAs are used regularly by major corporations. Under its flagship sustainability banner Ambition 2039, the German automotive leader Mercedes-Benz has committed to becoming a carbon-neutral company by 2039 along its entire value chain.[27] Hence, the firm is taking responsibility not only for emissions that occur during manufacturing but also for emissions that occur during the use phase. A huge proportion of the production-related emissions do not occur at Mercedes-Benz locations, but rather in its supply chain. Achieving transparency about emissions hotspots in the supply chain and quantifying possible reduction measures are therefore crucial, leading to Mercedes-Benz hiring Sphera and its detailed LCA models and supply chain knowledge to fill this gap.

For car manufacturers, the supply chain of the major materials is crucial. While working with Mercedes-Benz, Sphera conducted detailed CO_2 modeling of the supply chains for steel, aluminum, and plastics and derived and quantified possible reduction measures. This involved alternative production routes for steel, including via direct-reduced iron; using renewable energy in the supply chain of aluminum and plastics; and the use of secondary and bio-based materials. Wherever needed and relevant, additional environmental impacts such as acidification or water consumption were considered.

"The LCA models helped us gain transparency in our supply chain," says Dr. Klaus Ruhland, head of sustainability, corporate environmental protection, and energy management at Mercedes-Benz. "Quantifying decarbonization pathways for our main materials was the missing piece of our corporate decarbonization strategy."[28]

And it's not just corporations that make good use of cradle-to-grave LCAs. Take single-use plastic bags (SUPBs), for example, which are relied on heavily by consumers around the world. Significant debate has taken place regarding their environmental impact.

To identify which solution is most environmentally sustainable, the United Nations' Life Cycle Initiative compared the impact of SUPBs to their alternatives using LCAs that stretched from raw material extraction, production, logistics, and distribution to use and end of life. While SUPBs are commonly seen as the villain, this research showed that alternatives such as cotton and paper bags also produce an environmental footprint, and sometimes score worse than SUPBs in certain environmental categories.[29]

The study found that the SUPB is "a poor option in terms of litter on land, marine litter and microplastics, but it scores well in other environmental impact categories, such as climate change, acidification, eutrophication, water use and land use." Also, reusable bags can be environmentally superior to

SUPBs, but only if they are reused many times. For example, "a cotton bag needs to be used 50-150 times to have less impact on the climate compared to one SUPB."

The researchers concluded: "Reducing environmental impacts of bags is not just about choosing, banning, recommending or prescribing specific materials or bags, but also about changing consumer behavior to increase the reuse rate and to avoid littering. The shopping bag that has the least impact on the environment is the bag the consumer already has at home."

How Far Should an LCA Reach?

Supply chain partners may need to conduct their own LCAs to secure business with manufacturers. Therefore, close communication is needed to fully understand what kind of information is needed and why. Providing accurate, high-quality data can prove challenging for smaller companies, especially if they have never needed to measure operations to that level of detail. It's important to be consistent in data and methods, collecting data that is as temporally and geographically specific as possible while documenting all assumptions and justifications.

Data is often not available to the depth that is ideal for a robust LCA. Also, collecting appropriate data may take more time than some companies actually have available. LCA software and databases can help in guiding and facilitating the process, while sustainability software experts will help in terms of design and implementation. Major manufacturers may offer their own expertise to help suppliers and assets reach the necessary ratings.

After the impact results are gathered and formulated, they must be interpreted and shared clearly in terms of the previously decided goal and scope of the LCA. Results should be communicated in an authentic, credible, and easy-to-understand way. In addition to conducting the LCA, if more validation is requested, a supplier may even need to complete a critical review of the LCA in accordance with its own ISO 14040 framework. The diagram on the next page shows the four main steps in developing an LCA.

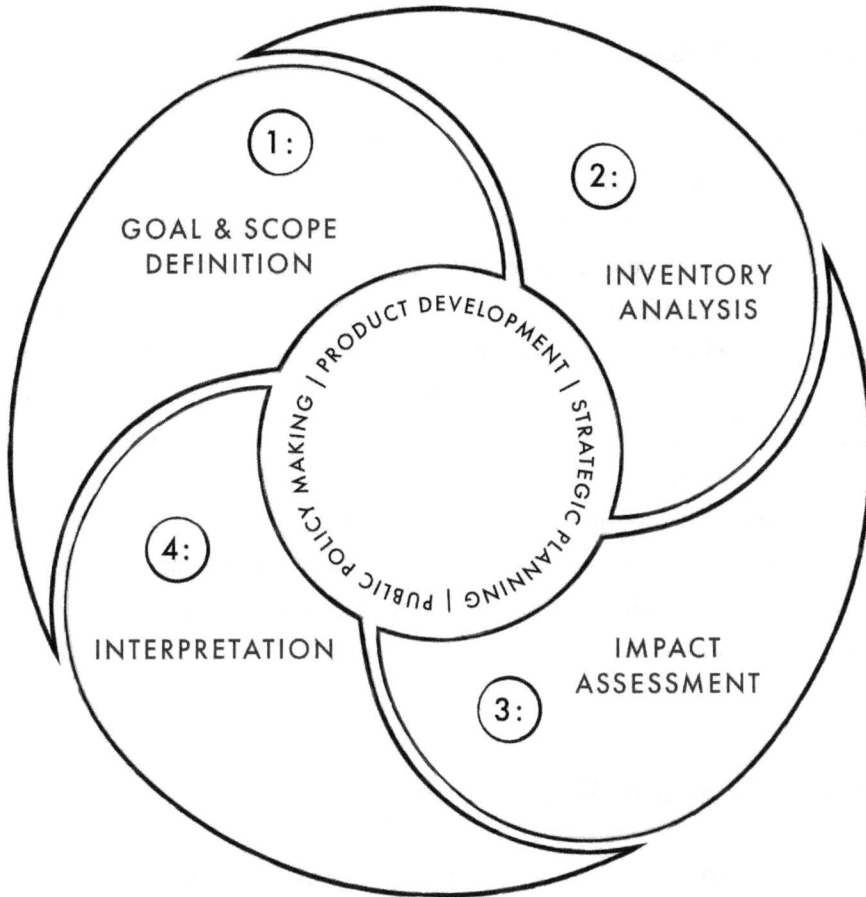

STEPS IN DEVELOPING A LIFE CYCLE ASSESSMENT

LCAs are notably used to influence the design of products. Of course, the sooner you apply an LCA in the product development process, the better chance you have of reducing impacts and improving process efficiencies. The greater your breadth of knowledge of emissions and resources from cradle to grave, the greater the opportunity of finding the optimal design on a cost and environmental or societal level. Cost and sustainability often run together.

LCAs don't directly calculate value in dollars, but it isn't hard to translate savings in energy and materials, or in reduced staff turnover, into actual costs.

LCA modeling can be a complex process that requires solid methodological expertise and technical versatility, but it's worth the effort, as the benefits will touch different departments across your enterprise. Product designers can incorporate LCA insights into product development to create innovative, more sustainable products. Sales teams can use the results to provide science-based evidence of products' environmental strengths. Marketing executives can back up green claims with solid facts and figures. With the right sustainability tools, an entire organization can quickly understand and leverage the results of sophisticated LCA studies to improve individual products, product lines, and the complete portfolio.

LCAs demand rigor. They can certainly cause a degree of positive disruption in the business. But they don't need to break the bank, especially with a targeted approach that investigates the most pertinent issues related to sustainability. The results are valuable.

Keep the Future in Your Hands

Meaningful progress in sustainability can take time. Comprehensive, accurate, and consistent ratings require extremely high-quality data, based on science and fact, and available in integrated software with prescriptive standards. Many firms are falling short of effectively acquiring and accurately utilizing data across the supply chain, leading to confusion and inaccuracy in reporting.

Firms that rest on their laurels while awaiting reporting frameworks to guide them do so at great risk. Customers, employees, and investors are running out of patience. Whether executives want to change or not, market forces and regulations will compel them to disclose their products' environmental footprint and societal impact. Being reticent will likely result in an unfavorable comparison to competitive brands and unfavorable response levels against consumer expectations.

No business should go it alone. It's far better to evolve and thrive as a cooperative contributing partner in your ecosystem than fall victim to natural selection.

Key Takeaways

- All parties in a business ecosystem can benefit from cooperation—and coopetition.
- Future-focused companies recognize that sustainability issues are really business issues . . .
- . . . And that strategic decisions on sustainability are a bet on the future.
- Companies will strengthen their reputations when sustainability drives innovation.
- Life cycle assessments (LCAs) can help reveal the complexities of your business ecosystem and its many opportunities.

8

SUPPLY CHAINS FOR GOOD

- It's supply chain risk *management*, not risk monitoring
- Sustainability adds a second dimension to supply chain risk management
- Visibility, automation, and control turn supply chains into a Force for Good
- AI-driven platforms can ensure timely, relevant supply chain alerts
- Scope 3 is both a challenge and a competitive opportunity

"I speak and speak . . . but the listener retains only the words he is expecting.
. . . It is not the voice that commands the story: it is the ear."
—Marco Polo, 13th century merchant and writer

As we discussed in the last chapter, Force-for-Good companies reach out into their ecosystems. Gaining a greater understanding of supply chains is an important part of that open-minded approach. But having provided all the valuable information generated by analyses such as life cycle assessments, what if it isn't good news? The next step is to play a hands-on role in managing the supply chain risks that emerge.

Supply chain risk management (SCRM) is exactly what it says—*management*. It is an active, not passive, discipline that requires constant attention for it to be impactful. As we'll discuss later, technology is critical in SCRM, but it's not the whole solution by any means. Relationship managers are needed to drive behavioral change with business partners, with regular conversations to convince suppliers that their resilience is your resilience and vice versa, and that SCRM represents good business for both. Creating and managing these relationships takes time and resources.

SCRM is a modern title for a concept that's as old as commerce itself. All the great merchant cultures in history understood the importance of protecting their trade routes. Empires and armies were only as strong as their supply chains.

The Industrial Age, with the arrival of the railway, extended the reach of supply chains from local to regional, national, and then international lines of communication. Through the 20th century, technological innovations such as warehousing, trucks, air freight, pallets and forklifts, and containerization combined to grow the volume and speed of goods transportation, which brought greater need for management of stock inventories, regulatory compliance, and quality control.

In 1967, IBM launched the first computerized inventory management and forecasting system, and soon information technology was tearing up the old paper-based systems and streamlining the logistics of supply chains beyond recognition. Barcodes and real-time warehouse management systems arrived in the 1970s, followed by a whole caravan of acronyms such as GPS, ERP, RFID, LCA, AI, ML, and IoT, each designating a technological innovation that helps to make supply chains more globalized and cost-efficient.[1]

Yet perhaps the most transformative collection of letters in the evolution of SCRM will prove to be PPP (People, Planet, Performance). Until recently, the primary function of SCRM was to ensure delivery that is on time, meets the required quantity, and maintains quality, all at the lowest cost possible. As long as raw materials and parts arrived on spec at the factory gate, then the backstory of how they got there was low on the list of priorities. Rapid globalization only made the provenance of goods that much more opaque.

Sustainability—and our improved ability to measure and track it—has introduced a new dimension for managing supply chains. As a result, the potential and, indeed, the requirement for enterprises to work toward a cleaner, safer, more equitable world has increased dramatically. Before, value creation happened mostly within the four walls of the business. Today, that is a distant memory, with global value chains of large companies spanning massive and complex networks.

Scaling the Agenda

The roots of today's SCRM also lie in financial health tracking through credit rating agencies. These risk profiles were set up to better understand the financial health of business partners at the moment of a sourcing decision. Annual updates of credit rating systems have tracked how financial health risk has evolved over decades.

In the early 2010s, the financial risk perspective in supply chain management was sharpened by several major disruptions caused by natural hazards and man-made disasters. In 2011, the Fukushima Daiichi Nuclear Power Plant in Japan was damaged by an earthquake and its resulting 45-foot-high tsunami, which claimed over 19,000 lives and disrupted the global supply chain for semiconductors. Just a few months later, devastating floods in Thailand caused $46.6 billion of economic losses, including massive disturbances to the automotive and personal computing industries. At the very end of the decade, the unprecedented and largely unimaginable disruption of the Covid-19 era revealed widespread weaknesses in supply chains, especially in areas such as medical equipment and pharmaceutical materials.

These massive supply chain disruptions have occurred against a backdrop of increased pressure from investors, talent, government regulation, communities, customers, and consumers for businesses to protect their sustainability profiles. Supply chain sustainability is an important tool in the box. Or, to put it in more positive terms, those enterprises that can create and demonstrate sustainable supply chains stand to attract more investors, retain the best people, gain a license to operate, and grow their long-term resilience. Through this lens, supply chain sustainability is a pillar of brand protection.

In the past, many suppliers of unbranded components—such as lower-tier manufacturers in the mobility sector—could always operate beneath the radar, as their names were never seen in the finished Audi, Airbus, or Alstom. However, they too now face competitive pressure to meet the expectations of their customers and share information, especially around emissions.

Sustainability has brought in a new realm of threat exposure to the supply chain that is increasingly powerful. Damage to brand and reputation should now be treated with the same importance as so-called force majeure threats, such as weather events or political instability. The business disruption from sustainability issues can prove longer lasting if companies or suppliers are sanctioned due to unethical behavior or banned from the market by regulators. These intangible threats bring very real consequences.

Visibility, Automation, Control (VAC)

To grapple with these increasing pressures, how should businesses go about achieving supply chain sustainability? By introducing measures for visibility, automation, and control (VAC)—including overlaps among all three—businesses can identify, monitor and respond to potential flare-ups in their supply chains. More than damage limitation, VAC can help businesses to mobilize their supply chain as a Force for Good.

Let's consider each element in the VAC framework, looking at the *what, how,* and *why* for each, as outlined in the table on the next page.

Elements of the VAC Framework	
Visibility	*What:* Visibility helps to drive risk awareness, improve your preventive capabilities, and enable the fastest crisis response possible.
	How: Get to know the players and structure of the end-to-end supply chain, which helps you understand your full threat exposure and the potential impact of those threats.
	Why: Without clear visibility, businesses view the world through a porthole rather than up high from the crow's nest, severely limiting the accuracy and completeness of their vision.
Automation	*What:* Automation empowers teams to react to risks faster and install preventive measures.
	How: Establish fully automated risk monitoring, ensure relevance of alerts, and develop improvement and mitigation frameworks.
	Why: Without effective automation, businesses may fail to access the right information or become overwhelmed by the wrong information.
Control	*What:* Control is the guardrail for full program adoption.
	How: Involve cross-functional and external stakeholders; ensure management buy-in, especially for trade-offs; and document, execute, and improve governance.
	Why: Without robust control measures, visibility and automation are likely to fail.

Visibility on a Need-to-Know Basis

The key to driving visibility is understanding the interdependencies within a supply chain, from end to end. What's the journey of materials or services? Which suppliers transport goods through which seaports, airports, warehouses, and distribution centers? Who are the business partners of your business partners, down the different tiers?

Understanding the full threat exposure in a supply chain is hard to do well in-house. Some enterprises have built their own "control towers" through which they connect supplier performance scores, financial health ratings, and prequalification questionnaires. They gather all known information within the organization to populate a digital twin of the supply chain. Those large enterprises with contingent business interruption insurance will usually get access to the leading natural hazard databases. Based on the geocodes for a given supplier site, they can connect to the risk exposure for storms, floods, earthquakes, and other specific threats.

That means a lot of data gathering, analysis, and decision-making about what is relevant or not. It's a slog. Even then, there is likely to be a lot of unknown information. That's why many enterprises choose to partner with external SCRM platforms, which have ready-made relationships with global data sources, such as cyber risk scoring providers, rating agencies, World Bank data, the CIA World Factbook, and plenty of others, all of which help to reveal latent risk exposure. In addition, they will do all the constant live web monitoring to flag incidents around the world that could potentially impact supply chain operations.

Coerce, Incentivize, or Team Up?

The next challenge is gathering information from further down the supply chain. Suppliers are inherently reluctant to share information, let alone to make changes in their own policies and practices, as there is nearly always a cost implication. Asking politely will usually lead to a polite rejection. When there is an equal power balance between buyer and seller, requests for data typically succeed at a rate of 5 to 10 percent. How then to push that rate to 80 percent and above?

Heiko Schwarz founded one of the leading SCRM software platforms in 2013 and is now a global supply chain risk advisor at Sphera. He observes:

> It's worth remembering that SCRM is not an isolated discipline. Leaders connect the collaboration of suppliers within sub-tier visibility programs with other decisions and levers, especially within procurement, such as supplier segmentation and rating practices. If suppliers want to be A-rated and enjoy long-term contracts, guaranteed volumes, and privileges, then they need to contribute to visibility programs.

Companies can underline incentives and penalties related to transparency in the request for information stages of a supplier relationship to increase leverage. Contracts are only awarded when the necessary questions are answered. These are the moments in the supplier life cycle when manufacturers can exert maximum pressure over a shortlist of final bidders to connect the dots and drive visibility over time toward 100 percent. Given that contracts may last the best part of a decade, a concession on visibility is a small price to pay for winning long-term work. By contrast, if the buyer asks for information in year six of a seven-year contract, then the leverage force is considerably weakened.

Schwarz offers further advice:

> A carrot can replace the stick, incentivizing suppliers to share information, by giving their data a business value. For example, those businesses that partner with SCRM platforms can offer access to suppliers, who can then receive very valuable risk insights from further up or down the chain, free of charge, which makes them a more resilient company too. Cascading information in this way enhances visibility across the entire supply chain.

It's not always the suppliers that put the supply network under stress. It might be the suppliers of the suppliers or even their suppliers. And understanding the structure, the dependencies, understanding where the spider in the web is, which might put the entire system under stress, is helping enterprises to be better informed. They then must team up to find ways to bypass a supply shortage together and to collaborate with business partners to mitigate systematic and systemic risks.

As these comments suggest, buyers need to embrace suppliers as part of the solution to drive more transparency within the entire network down the different tiers. As the term says, the supply network is a chain—and when any chain is under stress, it breaks at the weakest link.

Criticality Status

Understanding the exposure to threats is only truly useful when companies understand the potential impact of those threats. Yes, it's possible that anyone could be hit by a meteor as they walk down the street. But the chances are so infinitesimally small that nobody wears a crash helmet as a precaution.

This is the second most important dimension of risk management: combining threat with impact, known technically as *likelihood of occurrence* or *criticality.* Where are the elements within the supply chain that are more or less threatened? What would be the impact on the business if that risk materializes?

Take a crisis situation, such as a pandemic lockdown or a weather emergency. Let's say that within a matter of days, 25 of your suppliers in China are affected. This information is certainly important. But the next question is perhaps even more so: Which of those 25 has the biggest impact on your top and bottom line? Firefighting problems with 25 supply chains at the same time isn't feasible. So where should you prioritize?

Having a relocation timeframe in place ahead of time will pay dividends. If one site is knocked out by floods, how long will it take them to fix the problem and become operational again? If needed, how quickly could the business get machinery out and then ramp up production elsewhere? What are the stock levels required to overcome a disruption for a day, for 30 days, or for three months? Answers to all these questions must be paired with the level of threat exposure to make better business decisions.

While businesses can't influence the likelihood of a typhoon hitting a supplier, they can impose flood protection measures. They can drip-feed other suppliers with tiny orders as a backup strategy. Within the contract, they can insist on stock prioritization, so that their recovery timeline is faster than their competitors' (something that Japanese car manufacturers have made standard practice in recent decades).

Automation: Key to Monitoring Risk

In today's vast, deeply interconnected world, monitoring threats globally is a vast job. It's no exaggeration to say that large companies could employ 50,000 people to read newspapers around the world and trawl websites manually for relevant supply chain information—and the results still wouldn't compete with those generated by the best digital technology. For a start, SCRM programs can run new checks every second. Risk doesn't sleep, so this advantage allows companies to understand continuously what's changing in terms of threat exposure.

However, tech isn't a panacea. Applications need to be calibrated to ensure they don't *alert too much* or *alert too little*. The former results in a constant barrage of irrelevant alerts—distracting noise—that may ultimately cause the recipient to block out all notifications. This is called the Crying Wolf scenario, after the Aesop fable about the shepherd whose repeated false warnings about a wolf attack led village folk to ignore him—even when his warning eventually proved to be real. In the modern version, the irrelevant email sender is blocked, and all alerts are sent to the trash—including the critical ones about a fire in a supplier's factory or a border blockade.

Likewise, an alert level programmed to alert too little will also fail. In 2016, the border between Mexico and Texas was closed for security reasons due to a visit by the Pope that attracted an audience of 100,000 worshippers. Trucks were left stuck at the border, as analysts had failed to train their risk assessment model to recognize the Pope's visit as a supply chain risk. Afterward, the model was duly updated to reflect the capacity for disruption due to Papal visits.

This is where automation earns its dues. AI and machine learning can crunch through vast amounts of data and pick out only the most relevant snippets in terms of risk for the supply chain. Smart automation cuts out the noise in those billion data sets that need to be screened every second. The precision of AI reduces the risk insights to a few highly relevant indications that identify evolving threats before the risk materializes or (if not predictable) informs the customer faster than anyone else. Some SCRM platforms, such as Sphera's Supply Chain Risk Management application, have trained their AI systems for almost a decade, and their powers of deduction are frankly astounding. The software can also provide laser-sharp distribution of risk information to the right audiences across departments, reducing the time wasted on manual investigations, which can then be dedicated to prevention and mitigation.

Schwarz describes the impact of AI on SCRM:

> AI has turned financial health profiling from an annual credit rating into a real-time dashboard. Is the CFO leaving the company? Are there issues with regulators and penalties or even market exclusions that will lead to significant revenue losses for this business partner? Are their product launches delayed? Did they lose patents? Have there been sites closed or business units spun off or sold to private equity? All of these indications might be not critical because they are part of doing business, but technology can sense if they come in a

context within a specific timeframe of maybe one or two years, and all together, raising a red flag in terms of financial health.

This is really changing the game if you start to combine those hard facts coming out of a balance sheet, out of a profit and loss statement, together with those soft sensors that give you an advantage in terms of early warning signals to be much better prepared. Think about the rating scale for financial health next to a cyber risk score next to an insurance company's calculations of earthquake probability. I mean, there is not a single person out there that understands all those different scales and can interpret them, but technology can harmonize all this information and keep it up to date all the time.

It's worth vetting SCRM platforms to see which vendor works best for the business. They can all look the same when responding to a request for information, but it pays to run a pilot in parallel to check whether the noise cancellation layer is surfacing the most urgent information or missing the approaching wolves. Companies can recruit a new department to do prevention and mitigation for a more resilient supply chain, but they could end up devoting excess time to appraising whether an alert is useful or not. Risk management is then frozen under an avalanche of information.

Furthermore, all this tech wizardry is rendered comparatively useless if there aren't human systems to back it up. There needs to be a mandated role within the organization to drive prevention measures or rapid response to alerts, with a designated action flow across personnel, rather than a free-for-all. When the system is working well, these precautions can feel like an excessive cost—just like a virus scanner or a health insurance policy—but they are invaluable in times of need and can drive competitive advantage in a crisis situation.

It's also worth noting that every lesson learned should be fed into company standard operating procedures in due course. There should be a single point in the enterprise where the best-practice mitigation steps are documented and stored so that lessons learned today are operational tomorrow.

Control: All Joined Up

Staying in control of SCRM requires robust governance. Enterprises need to make sure that their definition of risk tolerance is tied to the company's strategy. For price leaders in apparel, for example, mitigation programs look very different from those of a surgical device company, which could lose its reputation in perpetuity through even the smallest quality issue.

All stakeholders within the enterprise need to acknowledge the risks that the SCRM team is scouting out. This should be done before the program is rolled out, as it's always harder to convince a manager who hasn't been involved from the start that they now need to make a budgetary contribution or provide resources. Don't neglect management buy-in either, because there will be trade-offs. There always are.

Under pressure from quarterly targets, buyers naturally gravitate toward the cheapest supplier. It therefore often takes senior intervention to say that the higher-quality option—which costs 20 percent more—is the preferred choice. Decisions made in line with the Noble Purpose of building a more resilient enterprise, and contributing to a safer and more sustainable world, will almost inevitably come with a short-term hit. Proper controls in your decision-making system can help insure that the right choices are made.

"This Is Not a Drill!"

Even with the most sophisticated SCRM programs, companies will still face disruptions. It's part of business. Leading organizations perform occasional "fire drill" exercises in which they simulate a crisis and invite their teams to work on the mitigation. In this way, people are trained in optimizing the crisis response. When the siren sounds for a *real* emergency, everybody executes their role through muscle memory, rather than giving way blind panic.

The control element of VAC also demands that everything be documented and that successes and failures are tracked. There needs to be a learning loop within the methodology to make sure that lessons learned get fed into the practices to continuously improve the program.

The greatest mistake is to believe that technology alone will solve the problem. That's why it's called risk *management*, not risk watching or monitoring. Without governance and the right mandates in place, programs are set up for failure, even though the design and technology look perfect on paper.

Manage Before Damage

Joyson Safety Systems (JSS) is a global leader in mobility safety, providing safety-critical components, systems, and technology to automotive and non-automotive markets. Its solutions allow its customers in the automotive sector the design freedom and confidence to drive the next generation in mobility.

To keep its supply chains running reliably, JSS needs accurate, real-time risk data and a single source of supply chain risk information. In the automotive industry, strict sets of government regulations and industry standards apply to ensure product safety, quality, reliability, and business integrity. As a tier-1 supplier of safety systems to automotive manufacturers, JSS manages sourcing processes in which compliance with such laws and rules is a top priority.

In addition, car makers are well known for their just-in-time manufacturing. Against the tightest of schedules, every tiny disruption costs a lot of money in downtime—estimated at around $250,000 per hour—and that's before any litigation to retrieve lost earnings.[2] As the Covid-19 pandemic revealed, this reliance on just-in-time sourcing makes automotive supply chains particularly vulnerable to disruption.

For all these reasons, the financial stability of suppliers has become a major risk for JSS, which relies on many small- to mid-sized suppliers for specialized parts, components, and materials. Recognizing the increased likelihood of suppliers becoming financially stressed and the need to stay ahead of possible bankruptcies, JSS has taken a preventive approach, avoiding the last-minute problem solving that had become commonplace in the past.

Ensuring the compliance and financial stability of its suppliers are two key reasons why JSS turned to a SCRM partner for support. The partner's platform was embedded into JSS systems, supporting functions such as critical parts management, dedicated crisis management, central supplier management, and topic management.

Now JSS can map suppliers, its own sites, supply paths, and further risk objects such as transport hubs on an interactive world map. Early warnings in near-real time enable crisis prevention, quicker adjustment in critical situations, and improved risk management capabilities, including:

- Comprehensive transparency and a total view of risk within JSS's value creation, including but not limited to supplier financial distress, geopolitical situations, natural hazards, reputational risks, and man-made events such as labor disputes and explosions.
- Real-time visibility on all supply chain risks. All data is embedded into source-to-pay procurement software within holistic supplier life cycle management processes. This enables JSS to combine centralized risk management with decentralized crisis management.
- Risk prevention for better sourcing and awarding decisions, along with proactive supplier development, allowing JSS to avoid conducting business with risky suppliers. Used in connection with crisis identification, JSS can react as one of the first on the market, enabling it to fulfill the company maxim "manage before damage."

Case in point: When one of JSS's suppliers in Latin America burned down completely, reaction time was critical. Alerted by the new SCRM system, JSS could immediately send a technician to what was left of the supplier's facility. JSS was able to retrieve its specialized tools, minimally damaged, from the ashes. The company refurbished its tools and shipped them to a partner who could continue production.

Left to its own devices, the supplier likely wouldn't have informed JSS of the fire until a few days later—a pivotal period, as JSS had limited inventory levels. Instead, they managed to avoid production downtime. The company can now mitigate any risk in its supply chain by almost 100 percent and improve the productivity of its purchasing team by more than 60 percent.

Seven Tips to Get Ahead in SCRM

1. Ensure supply chain risk management is linked to other activities, especially procurement, to maximize leverage when seeking information or behavioral change.
2. As part of the risk inventory, ask stakeholders what they perceive as risk, as this encourages their contribution when support is needed. Mitigation is not a silent activity.
3. Prioritize management buy-in from the start. There will be trade-offs.
4. Always start with the suppliers that matter most, then widen the coverage using the lessons learned.
5. Remember that suppliers managed poorly are part of the problem; if managed well, they are part of the solution.
6. Good programs continuously educate and evangelize internally why risk is an important part of business decisions. Risk shouldn't be suffocating—there are no risk-free decisions—but it always needs to be considered.
7. Embrace change as an opportunity, especially in the field of regulation. Strive for continuous improvement.

Staying Within Scope with Supply Chain Sustainability

In previous chapters, we have touched on the growing urgency of Scope 3 responsibilities, which demand that companies provide information—

especially carbon emissions data—from across their entire supply chain. Value chain emissions can account for up to 80 percent of a consumer company's overall environmental impact and more than 90 percent of "the impact on air, land, water, biodiversity and geological resources," according to a report by McKinsey.[3] Suppliers therefore play an outsized role in a science-based decarbonization strategy to achieve net zero.

Scope 3 is both a threat and an opportunity, depending on how well companies respond to the challenge. As part of commitments to achieve net zero or halve emissions by 2030, 2040, or 2050, companies must prove to bank investors, their boards, employees, and customers that they can measure the carbon emissions within their own four walls—but also along the supply chain using a science-based approach that is immune to accusations of greenwashing.

Without the contribution of suppliers, at least in terms of information provision, this simply isn't going to happen. Buyers must therefore work on the process of building and advancing relations with suppliers, who have the advantage, especially as reduction measures usually come at a cost.

Reliable supply chain platforms are foundational to building sustainable, long-term relations with suppliers and are a useful lever for buyers. They can help companies both to assess the current carbon footprint of suppliers and to collaborate with suppliers to reduce their combined carbon footprint. Typical SCRM platform capabilities include suppliers, so they support the entire range of required activities, including supplier risk assessments, supplier network collaborations, monitoring exceptions, and mitigation activities that support cross-functional collaboration, both internal and external.

Over time, SCRM programs that extend their focus to environmental and societal risks will evolve into supply chain sustainability platforms. Again, working with a third-party supply chain sustainability platform that maintains business relations with all suppliers through its platform is a prudent way of streamlining Scope 3.

For the last decade, Alex Gershenson has been a pioneer in the supply chain sustainability space in founding innovative software companies in the market. Gershenson has witnessed the sustainability movement gather momentum on the front lines. He observes:

> In order to fulfill your commitments to your shareholders, to your employees, to your consumers, and to the regulators, you have to account not just for the things that happen within your four walls, literal or figurative. You have to account for your total footprint, whether it be plastics, deforestation, carbon, waste, energy use, you name it. Without looking into your supply chain, you can't honestly report on your commitments, and you can't make meaningful change. The supply chain is therefore the area of maximum impact when it comes to making a difference on sustainability.

Gershenson sees plenty of desire among businesses to meet their Planet and People targets, but the capability is yet to catch up with the ambition:

> It's worth remembering that everybody is on a journey toward their goals. Some companies are further up the gradient than others. There needs to be a degree of pragmatism. If you hit your suppliers with a product-level data request as your first outreach around sustainability, that approach is going to fail, because they won't have the maturity to answer that. Better to start at a company-level assessment, engaging with suppliers and saying: Do you track this? Do you track that? Do you have a policy on this?

Unlike other kinds of compliance, where you can submit your audit results or financial statements for the quarter, sustainability isn't there yet. "It will be," Gershenson believes:

> In the next five to 10 years, companies will need to submit their carbon emissions, for example, and everybody will have them on hand. It will be standard. But not yet. We're in this fascinating transition period where everybody recognizes that this is a thing that they

must address, but they don't yet have the expertise or experience to fulfill all their goals.

It makes little sense for every company to walk slowly along the road to sustainability when software can propel many forward together at a far greater speed. When buyers, suppliers, technology providers, and data sources partner up to form supply chain sustainability platforms, then company decision-makers can start to see the whole picture and also the hotspots that need to be addressed first. They can see where to get the biggest return on their sustainability investment. Gershenson notes:

> People make better decisions with better information. Supply chain sustainability platforms need to create transparency in supply chains so that decision-makers can champion sustainable procurement based on the information that they have. The software enables data sharing between buyers and suppliers to ensure rapid transition of information and action to manage risk.

By joining a supply chain sustainability platform, companies can see the status and results of suppliers, and their suppliers, and so on to the Nth tier. They can navigate complex global supply chains and evolving regulations; easily share their data with suppliers, internal teams, and their customers; collaborate with suppliers to incentivize improvement; and through all of this, they can accelerate progress toward their goals. As more regulations are passed that demand transparency, this kind of solution will become indispensable in meeting global regulatory requirements and stakeholder expectations.

Holding the Right Conversations

SCRM software connects buyers and suppliers in one unified platform that enables responsible sourcing and emissions tracking, sharpens environmental and societal risk detection, and elevates sustainability reporting.

Let's say that a U.S. novelty retailer sources hand-painted wooden Christmas decorations from a firm in Germany. This supplier has a factory in Frankfurt and a factory in Sri Lanka. Using the platform, the retailer can see where both factories source their wood (Sweden and Indonesia), the labor standards, water and energy consumption, and so on. They can then make an informed decision on whether they want premium decorations made in Germany or more affordable ones from Sri Lanka.

Let's take it a step further. Perhaps the retailer prefers the Sri Lankan decorations, but they need the wood supplier in Indonesia to demonstrate they are certified by the Forest Stewardship Council and only source from responsible timber plantations, rather than native forests. They can have that conversation on the SCRM platform, either through their supplier or by going directly to the third-tier supplier in Indonesia.

Whatever the retailer decides, note the power of the SCRM platform. In the past, the retailer might have had to make a decision based on cost alone. Or they might have avoided the supplier altogether because some of their decorations were made outside Germany, which represented an unnecessary risk. The platform makes smarter, more informed choices much easier.

Whole industries are taking this matter in hand with a view to raising sustainability standards, disseminating information, and making it easier for members to do business. For example, The Sustainability Insight System (THESIS) Index is run by the Sustainability Consortium, the conglomeration of retailers, nonprofits, brands, and universities formed by Walmart.[4] The index is designed around product-level sustainability of items that are sold in major retailers around the world.

THESIS poses 15 questions in four broad categories (energy and climate; material efficiency; natural resources; and people and community) to 100,000 of its suppliers worldwide, many of which have needed to advance their own sustainability credentials to retain a seat at the table.

Through this assessment, THESIS helps brands and manufacturers understand the sustainability story of their products. Suppliers can see their own scores and understand how they compare to others in the field. This gives them insight into how they can improve in each of the categories they supply. To date, 70 percent of Walmart's goods come from suppliers that have participated in the index.

"So, instead of each company trying to figure what's material for their company, they can assess the *category* of the goods they sell," says Gershenson. "If you sell consumer electronics, you're going to be asked different questions by a seller of pineapples or pet mice. With THESIS, brands like Coke can respond once to all of their retailers that they sell goods through, which hugely reduces their burden as a supplier."

In addition, THESIS provides methodologies for businesses looking to move forward on their sustainability journey. Gershenson notes, "It's not just about reporting but getting resources to improve. By reducing emissions or waste, you're saving money. Members can also benchmark against other suppliers. And they gain recognition from their customers, as a return on their sustainability investment."

Businesses need to grasp and embrace the process of building sustainable supply chains. It will not happen overnight. It requires collaboration among you, your customers, your suppliers, and your internal stakeholders. It also requires an understanding of where you want to be and how you'll get there faster than your competitors.

Regarding supply chain sustainability, Gershenson is bullish about the pace of adoption:

> Within the next five years, any company that cannot say, "These are our carbon emissions from our supply chain, this is what we're doing, this is our plan, these are our targets, this is how we're tracking to them," . . . that company is gone. Game over.

Ten years ago, supply chain sustainability was a niche conversation. Nobody wanted to listen. People thought you were odd to mention it. Now, it is top of the agenda of every procurement organization. By 2028, if your business cannot credibly, with transparent and traceable data, point to your supply chain's environmental and social impacts, and show a mitigation plan, then you're going out of business. Period. If it's not regulatory issues or consumer boycotts, then it will be employee revolts or lack of investors.

Signify: Sophisticated Procurement

Based in the Netherlands, Signify is the market leader in the extremely competitive LED lighting systems and software market. To lead, their procurement teams must manage their time well, particularly when unexpected events occur, so they can concentrate on strategic tasks. In recent years, a series of disruptions, including Covid-19, shortages of key components, and soaring freight and energy costs, have put even greater demands on these teams.

Managing procurement risks takes time, effort, and constant innovation, using a robust procurement platform combined with real-time risk monitoring to onboard and manage the best suppliers. For example, if capacity is down because of a fire at a factory, Signify needs to find capacity elsewhere—and fast. They need to know which suppliers can provide the components quickly and in the required quantity.

The procurement team has recently integrated a supply chain sustainability platform and followed the VAC approach.[5] Signify can now assess new or existing customers on all types of risks, ranging from financial, regulatory, and geopolitical risk to natural hazards and man-made risks such as explosions, as well as reputational risks linked to sustainability.

And it does all this in real time, feeding risk scores into the solution, where they can inform processes such as sourcing and category management.

The department's skillset has advanced from spreadsheet management to big data analysis and visualization tools.

The procurement team is also breaking free from the silo mentality to ensure that data is shared, with reporting and risk data analysis made consistent across the organization. Procurement processes and reports are automated, leading to greater standardization, improved efficiency, and more accurate decision-making. Signify is now harnessing the power of AI to negotiate with its suppliers.

What's the Cost of Disruption . . . ?

The downsides of supply chain interruption are real. According to a survey by *The Economist,* since 2020, "disruptions have incurred substantial financial costs (averaging 6 to 10% of annual revenues), in addition to reputational costs—in terms of customer complaints and damage to brand reputation—as businesses struggled to maintain supplies of their goods."[6] Of course, it's worth noting that this period included the full brunt of Covid-19. Research from the McKinsey Global Institute calculated that "over the course of a decade, the average company can expect to lose nearly half of a year's profits from supply chain disruptions."[7] Just one lengthy disturbance to production could cost 30 to 50 percent of a year's EBIDTA, while a shorter interruption of 30 days or less could still equate to a 3 to 5 percent loss, the same study found.

Additional drawbacks such as customer churn, loss of market share, and buying from the competition also represent financial risks, as do penalties due to non-delivery, late delivery, or partial delivery. Companies might face consumer boycotts and regulatory penalties tied to revenue or market exclusions linked to non-compliance.

. . . And What Are the Financial Benefits of SCRM?

Here are some quick examples of customers finding a sizeable upside to implementing SCRM, drawn from across the Sphera network.

- *Avoiding unplanned costs.* Cost saving is a high priority in all companies, and it can be disheartening to see those savings eaten up by unscheduled expenses, such as chartering an airplane or helicopter to shuttle parts from A to B to avoid letting down a customer. This was the dilemma faced by one automotive supplier that was paying $35 million a year on emergency flights due to supply chain disruption. Within a year of introducing proactive supply chain risk management measures, the crisis trips had stopped and related expenses were reduced to zero.
- *Freeing up working capital.* Many companies need to maintain a large, costly inventory buffer to mitigate risk. The greater the risk visibility, the more risk-aware decisions are made, letting companies can reduce the amount of bound capital in the inventory buffer. One customer used SCRM support to lower their threat level. As a result, they had the confidence to reduce bound working capital by $26 million.
- *Saving on insurance.* Companies that take contingent business interruption insurance can also make savings through SCRM, just as a household insurance policy might be lowered by installing a fire alarm or intruder alert. One customer successfully reduced their premium fees by $2 million a year, as the insurer downgraded their risk profile.
- *Improving speed of response.* A customer in the chemicals sector was notified instantly when a supplier's refinery was set on fire. They could gauge immediately that the refinery would not return to operations within at least the next three months, so there was no choice but to buy the relevant chemicals at a higher price on the spot market. While this expenditure was painful, the company outmaneuvered the competition, who were forced to buy the same chemicals at a 20

percent higher price the next day, after the market had adjusted to the drop in supply. The customer saved $275,000.

These examples show how SCRM can help an enterprise avoid revenue losses and turn risk into a competitive advantage by delivering when rivals are stymied. Companies need to anticipate what's coming, be prepared to minimize the potential impact, and reduce the likelihood of a risk event materializing. Get that right, and SCRM will bring real value to the enterprise.

We live in a world full of risk. It's the new normal that disruptions happen, whether financial, health, cyber, natural hazards, conflict, or sustainability. Disruption has become a constant—which means we must be better prepared and switch from reactive to proactive.

That's when risk becomes a competitive advantage. By embracing the opportunity, risk, too, becomes a Force for Good.

Key Takeaways

- Visibility, automation, and control (VAC) should be your goals when planning to minimize risk.
- Supply chain risk management (SCRM) is about proactively acting to minimize risks, not simply monitoring them.
- Link SCRM to procurement to maximize your leverage with suppliers.
- SCRM platforms can help avoid firefighting costs, release bound capital, and lower insurance premiums.
- Speed of response from SCRM is a source of advantage.
- Supply chain sustainability connects buyers and suppliers in one unified platform, enabling responsible sourcing and emissions tracking, sharpening environmental and societal risk detection, and elevating sustainability reporting.

9

WHAT DOES GOOD

LOOK LIKE?

- Meet a fictional business that has adapted to change since 1882
- Its ability to pivot secured growth throughout the 20th century
- Business collapse in the 2010s was avoided by turning to enterprise sustainability management
- Now, the company is building for the future with optimism

"The decision we are making today is about securing our company for tomorrow. In fact, I see no other choice."
—Eliza Bourne, CEO of Furtilux Furniture

W e've explored how sustainability provides a platform for companies to become a Force for Good in today's world. But what does *good* actually look like? This final chapter provides a (fictional) case study of how a business (Furtilux Furniture) can respond to cultural change. Innovation can help build a stronger future for the courageous business—and for all the stakeholders that are touched by its influence.

Just for clarity, this is not an example of perfection in action. There's no such thing, in my view. Yet it does show that an optimal approach to operationalizing sustainability is within reach.

Good Fortune Favors the Brave

Furtilux Furniture is one of many successful American companies that were born during the rapid industrialization of the 1880s, created by the hard work and imagination of a solitary entrepreneur. But the company, founded in Chicago in 1882 by Frank Bourne, wasn't called Furtilux to begin with, and at the time it didn't even make furniture. This is its story.

Frank worked as a wheelwright for the carriage manufacturers Coan & Ten Broeke. For 12 hours a day, he turned wooden spokes for the fancy buggies that transported the burgeoning middle classes through the rutted streets from the factory gates to their uptown residences. Ambitious and bright, Frank wanted to ride at their level rather than tramp home through the mud. So he came up with a plan.

Workers like Frank were given an allocation of wooden offcuts and defective panels, hubs, and wheel rims to burn for warmth at home during the winter. Despite the subzero temperatures, Frank stockpiled his share. On his day off, he worked at a cooperage (a barrel-maker), asking only that they pay him in wooden staves. By the time spring arrived, Frank had enough spare parts to build two fine hitch wagons.

Instead of selling the carts, Frank took them to show his employers. Impressed, they allowed him to set up a workshop in an outbuilding, where he began producing wagons for use in Chicago's steel mills and meatpacking plants—all from offcuts of wood.

Demand grew fast, and Frank's sideline business swelled in response. But he was still walking to work through the mud. Having built up a team to replace him, Frank took out a sizeable bank loan and set up his own factory on the banks of the Calumet River, where he received a ready stream of wood for his Finest Bourne Wagons. His business flourished, soon needing as many hands as his previous employer to meet demand. Every morning and evening for 30 years, Frank traveled through the center of Chicago in his own fancy buggy. He died in 1910, just two days after paying off the final installment on his business loan.

From Carts to Cots

By the 1920s, cultural change had swept through Illinois in the form of the automobile. The factory was now run by Frank's two sons, Charles and William. Charles was a reluctant businessman. "If it ain't broke, don't fix it," was his motto. "There will always be a need for horse-drawn wagons."

But William could see which way the winds were blowing. In an extraordinary move for the time, he gathered feedback from various stakeholders, such as customers, employees, and potential future investors. Should they stick with carts or pivot toward wooden furniture? The votes were unanimously in favor of change—and Charles never spoke to his brother again.

Through William's direction and the loyalty of the workforce, the company survived the Great Depression. At the end of the Second World War, Bourne's range of bedsteads and nursery cots played an instrumental role in the Chicago baby boom, while suburban homemakers snapped up their tables, chairs, wardrobes, and kitchen dressers.

William had no children of his own, however, so when he died the business passed to his nephew David Bourne. Despite his father's opposition, David leaped at the opportunity. He was more Frank than Charles. In the 1970s, he expanded the factory in Chicago and opened up new facilities throughout the Midwest. David predicted the flood in imported, mass-produced furniture and pivoted the business again, putting nearly every egg in the family basket into hotel contracting (though he retained the company's showroom store on Michigan Avenue). To make the break clean, David changed the trading name to Furtilux Furniture.

The next two decades were a period of constant growth. Furtilux's goods could be found in hospitality venues from San Francisco to Miami, Toronto to Cancun. The *Wall Street Journal* described Furtilux as "the most comfortable brand you've never heard of."

In 1992, the company went public. By the time David died 10 years later, Furtilux could be found on the Fortune 1000 and had entered new markets in Europe. He had connected a vast network of manufacturers and suppliers of raw materials who imported wood, plastics, polyester, and natural fibers like wool and cotton from across the globe.

What Goes Up . . .

David had three children from two marriages: Chuck, Dave Jr., and Eliza, who arrived 10 years after her stepbrothers. The Bourne boys didn't care much for business or hard work, quickly shedding much of their stock to venture capitalists to finance their playboy lifestyles. Furtilux became embroiled in a race to the bottom on price, as the company tried to compete with mass producers for contracts with a new wave of budget hotel chains. Suppliers and workers were squeezed in the name of efficiency.

The post-2008 economic downturn revealed the cracks that were widening at Furtilux. A series of articles in the *Chicago Tribune* carried complaints of maltreatment by employees whose families had worked for Furtilux over many generations. Another front-page headline splashed historical allegations of chemical pollution in the Calumet River. Furtilux was implicated in a report by the *New York Times* on hotel furniture supply chains, including examples of forced labor in textile factories and unsustainable logging practices.

With Chuck in rehab and Dave Jr. serving a suspended sentence for tax evasion, Furtilux was looking for a buyer. But no good offers came. After 130 years, the company that Frank built was on the verge of collapse. Was the fourth generation of family leadership destined to be the last?

Phoenix from the Ashes

Enter Eliza, the only child from David's second marriage. Her mother, a teacher who had emigrated to the U.S. from the Philippines, insisted that Eliza reach her potential at high school. She went on to graduate summa cum laude in business administration from the University of Illinois, where she also obtained a certificate in sustainable enterprise.

Inspired by the resourcefulness of her great-grandfather, she founded a startup with a designer friend, making affordable furniture from recycled materials and upcycled goods that were destined for the landfill. Her company ReBourne was featured in *Wallpaper*'s "Young Brands to Watch" in 2009— just as Furtilux was hitting the skids.

Having scaled up and opened new stores in Detroit, Milwaukee, and Indianapolis, Eliza felt ready to offer her services to Furtilux. The business was in dire need of upcycling itself. She planned to transform the fortunes of the struggling firm by using enterprise sustainability management as a platform for change. As CEO, she would attract fresh investment from funds that were looking to back sustainable companies. For all its recent problems, she knew Furtilux still had an enviable heritage in the hospitality industry and extensive networks across the globe.

"The culture of society is changing," she wrote in her proposal:

> Hotels will increasingly choose suppliers based on their sustainability credentials. As my great-uncle William Bourne showed in the 1930s, people want to work for companies that care about their well-being and appreciate their contribution. If we act now, we move ahead of the curve. If we do nothing, and merely stumble forward as

we did before, then Furtilux will fall into terminal decline. There will be severe consequences for our supply chain, our employees, and the communities where we work. I don't want to see that happen. Because it doesn't need to happen.

On May 1st, 2015, Eliza Bourne was announced as the new CEO of Furtilux. The transformation had begun.

Follow the North Star

Eliza's first move was to diversify the board, bringing in various views from domestic and international constituencies. She created an office of sustainability led by a new chief sustainability officer (CSO), ensuring a clear line of sight between the CSO and finance director.

Eliza aligned sustainability to the company purpose, which now reads, "Guided by innovation, Furtilux is a net contributor to the economy, society, and the planet we share."

The company values were redrafted to focus on innovation, safety, collaboration, resilience, and pride, under the collective title of "What would Frank do?" The company tagline changed too, from "Better off with Furtilux" to "Furnishing our Future."

Before sharing the new purpose externally, Eliza announced the change privately to a select group of potential investors, workers, community representatives, and suppliers in a town hall meeting. She apologized for mistakes that had been made in the last decade, then set out the case and the road map for change:

> The decision we are making today is about securing our company for tomorrow. In fact, I see no other choice. Unless we turn Furtilux into a Force for Good in the world, then we will cease to exist. I won't sugarcoat our predicament. Our company is in the grip of a crisis. Jobs are on the line. There were errors of judgment, including many by

my own family. In the future, I will own any missteps I make, openly and willingly. But I will not stand idly by as this fine business is consigned to the history books. I will take action.

By using sustainability criteria as a basis for change, Furtilux will do more than provide our customers with what they seek. I believe we will meet our own needs for job satisfaction and company pride while also making a profitable return on our investments. We will do what's right for our communities and for the supply chains that depend on us. Believe me, it won't always be easy. There will be challenges and setbacks. But we will succeed if we take this journey together. I look forward to seeing how the new Furtilux looks and works—and also how it feels. Our future starts now.

In Order of Material Priority

As a next step, Furtilux partnered with a sustainability strategist to map out the transition journey, including a commitment to net-zero carbon emissions by 2050. The consultant used a software program to conduct a materiality assessment that surveyed the views of all stakeholders to lay bare the company baseline. Only then could the company commit to a clear path forward. By charting the results on a materiality matrix, Furtilux prioritized its next strategic steps to secure immediate business gains (next page).

The matrix revealed that Furtilux's customers—and especially the customers' customers—were demanding furniture made from more sustainable materials that had a lower carbon footprint. Since the exposé by the *New York Times*, hotels had stressed the need for transparency around supply chain provenance.

Water consumption, energy efficiency, and sustainable factory design were also highlighted as issues, although they scored lower on the priority scale.

In the social category, employees were broadly satisfied by the company's pay structures, but they lacked motivation in their work. Factory

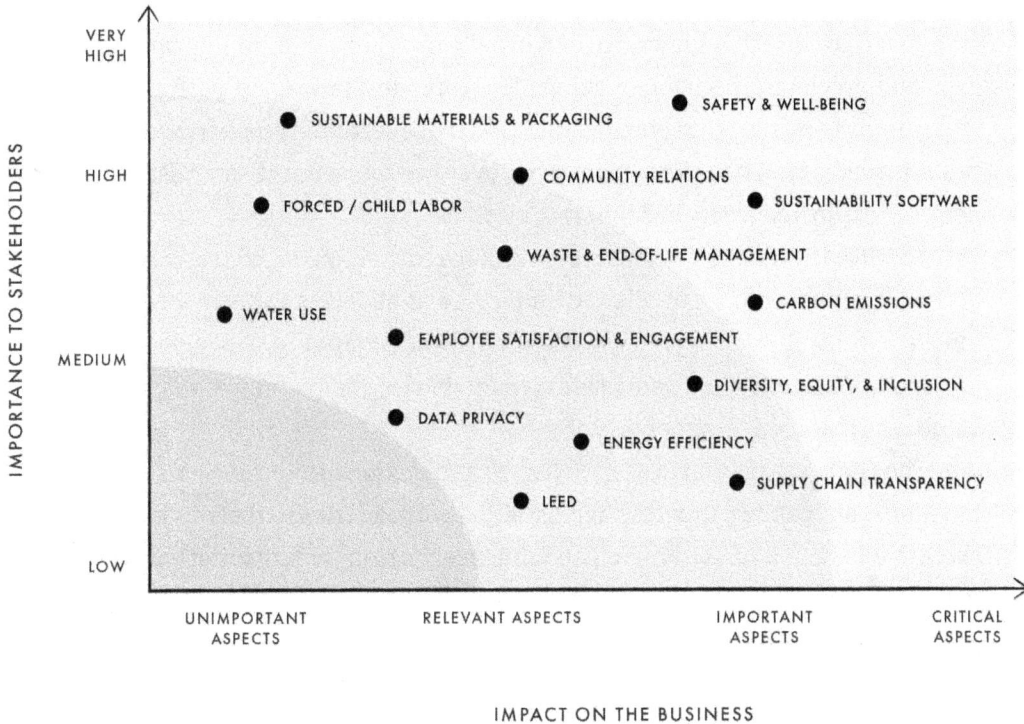

The chart is a scatter plot titled at the axes. The vertical axis is labeled "IMPORTANCE TO STAKEHOLDERS" with gradations LOW, MEDIUM, HIGH, VERY HIGH. The horizontal axis is labeled "IMPACT ON THE BUSINESS" with gradations UNIMPORTANT ASPECTS, RELEVANT ASPECTS, IMPORTANT ASPECTS, CRITICAL ASPECTS. Plotted points include: SUSTAINABLE MATERIALS & PACKAGING, SAFETY & WELL-BEING, COMMUNITY RELATIONS, FORCED / CHILD LABOR, SUSTAINABILITY SOFTWARE, WASTE & END-OF-LIFE MANAGEMENT, WATER USE, CARBON EMISSIONS, EMPLOYEE SATISFACTION & ENGAGEMENT, DIVERSITY, EQUITY, & INCLUSION, DATA PRIVACY, ENERGY EFFICIENCY, SUPPLY CHAIN TRANSPARENCY, LEED.

THE FURTILUX MATERIALITY MATRIX

workers also voiced urgent concerns about onsite safety. In Chicago, local community groups portrayed Furtilux as a business that had lost sight of its roots.

Responding to concerns around health and safety at work was the highest item on the to-do list. Eliza commissioned an investigation of current practices and found that the company lacked sufficient tools to proactively manage process safety. Factories were using years-old audit and inspection data to make critical health and safety decisions. Infrequent, incomplete inspections and outdated, siloed data were putting workers at risk.

Furtilux set about installing a better safety strategy using specialist software to anticipate and adapt to risk as it evolved. For example, the company centralized its incident reporting, audit management, and risk assessment systems to create a "single source of truth," which helped to fill information gaps and accelerate decision-making. The company saw a 500 percent increase in risk and hazard reporting. Initially, these seemingly negative statistics provided a shock to the system—but over the following 12 months, they helped Furtilux reduce its incident rate by 70 percent.

To win back and then strengthen workforce trust, Eliza opened clear channels of communication that allowed employees at all levels to share ideas for innovation as well as complaints, backed by a confidential whistle-blowing policy. Operational managers were made accountable for demonstrating sustainable change in their departments.

The CSO extended the use of software applications to include measurement of sustainability progress against KPIs, with metrics to incentivize the executive team and disseminate good practices across the organization. Not to be outdone, the chief human resources officer launched an employee engagement survey to measure current sentiment and was made accountable for increasing the score against his own KPIs. Improvements were communicated broadly, both internally and externally.

Expansion and Regulation

Next up were supply chain transparency and the use of sustainable materials. Eliza commissioned cradle-to-grave LCAs of every product in the Furtilux portfolio. Surprisingly, Furtilux had done little to understand the needs of hotel guests—the ultimate customers—for the last 40 years, and still worked to profiles of 1980s clientele.

As part of the LCAs, Furtilux calculated the product carbon footprint for goods that were shipped to Europe (often using materials that had already voyaged to the U.S. from Europe and Asia). The results showed that opening a new plant in Germany would prove more cost- and carbon-efficient. Further measurements suggested that acquiring a manufacturing firm would be better for both the economic and environmental bottom lines than building from scratch.

As a global company, Furtilux now faced new regulatory challenges, including European Union environmental standards around the use of chemicals, in addition to the German Supply Chain Act, which demanded higher levels of scrutiny and transparency. Within a few years, Eliza decreed that all of Furtilux's products in the U.S. should aspire to German manufacturing standards as a way of getting out in front of regulatory change. The tail had wagged the big dog.

These immediate governance concerns were alleviated by the use of supply chain risk management software, designed to help companies maintain business continuity by becoming more risk-aware, reacting faster, and managing against emerging disruptions and compliance requirements. Furtilux chose an AI-powered solution for improving preparedness across the entire organization.

The product LCAs relied on clear People, Planet, and Performance data from suppliers. Furtilux terminated accounts where suppliers represented an unacceptable risk by falling too far short of environmental and social standards. For other smaller suppliers, particularly in developing economies, Furtilux provided support to help capture and process the right sustainability data—and then make improvements where necessary. In 2023, Eliza decided that all the timber used in U.S. factories would be sourced from sustainable logging operations around the Great Lakes.

More recently, Furtilux has managed to tackle issues that scored lower down the materiality matrix, such as taking measures to reduce water and energy consumption. The company has set targets to lower the use of both by 30 percent by 2030. Factories are currently being upgraded to meet LEED green building standards.

People and Planet Contribute to High Performance

Furtilux partnered with initiatives such as UL GREENGUARD certification to ensure that all of its furniture met standards for airborne chemicals—commonly referred to as volatile organic compounds (VOCs)—which can cause headaches, eye, nose, and throat irritation, and dizziness. Long-term exposure to certain VOCs may lead to chronic diseases or cancer. Hotels that meet the GREENGUARD standard are increasingly attractive to parents with young children and those with respiratory conditions.

The company entered a corporate sustainability partnership with a leading hotel chain and joined its Good Neighbor Policy, a comprehensive community outreach program. Furtilux now donates used furniture to local organizations and shelters in each of the cities where the chain is present while encouraging volunteer activities among its own employees.

Despite misgivings from the finance department, Eliza insisted on setting a budget aside every year for R&D, with a focus on innovation around the circular economy. This proactive approach bore fruit in 2022, when Furtilux won a contract with a major global hotel group for a new chain of boutique hotels across Europe that used upcycled materials in its communal areas. Furtilux's foresight in establishing a base in Germany was critical for securing the deal.

Using research on nontoxic varnishes, Furtilux patented a durable chemical compound that is now a standalone product, creating a new million-dollar income stream. In partnership with a well-known manufacturer of office furniture, the company formed a Design for Inclusivity initiative that invited people with disabilities to take part in the development of more accessible products.

Having grown its reputation as a sustainability leader, Furtilux began working with hotel chains as an advocate for behavioral change. For example, a range of restaurant furniture made from recycled wood carried a QR code that not only provided guests with a smartphone menu, but also information about the provenance of their tables and chairs and how to recycle their own furniture and mattresses at home.

In 2022, Eliza proudly opened a factory seconds store in downtown Chicago on the same day as an interactive provenance zone in its Michigan Avenue showroom, where customers can take a virtual tour of supply chains to see where materials are sourced.

Inspired by the Swedish manufacturer IKEA, Furtilux works with upcycling influencers to broadcast furniture hacks on its popular YouTube channel. Chuck and Dave Jr. have returned to the family nest and now host a monthly podcast on eco-fabrics.

Measuring the Results of Enterprise Sustainability Management

From a brand perspective, Furtilux has recovered its reputation on the back of an aggressive marketing campaign. Under the headline "Look who's back in the green," the *Chicago Tribune* ran a glowing editorial that praised the transformation.

Since 2015, when Eliza took over as CEO, KPIs for employee motivation and pride have shown an increase from 35 percent to 85 percent. Staff churn has dropped from 28 percent annually to less than 10 percent. During the Covid-19 pandemic, Furtilux was able to retain 99 percent of its workforce, despite reducing wages at the peak of lockdown.

Within several years, Furtilux had the funds to offer greater employee support, with a full package of benefits including life insurance, healthcare, disability coverage, parental leave, retirement provisions, and also stock ownership.

Most tellingly, Furtilux's share price has recovered to pre-2000 levels, with a healthy growth forecast for the rest of the decade (next page). It's just as Frank would have wanted. Except that Eliza takes the Chicago elevated electric train ("the L") to work.

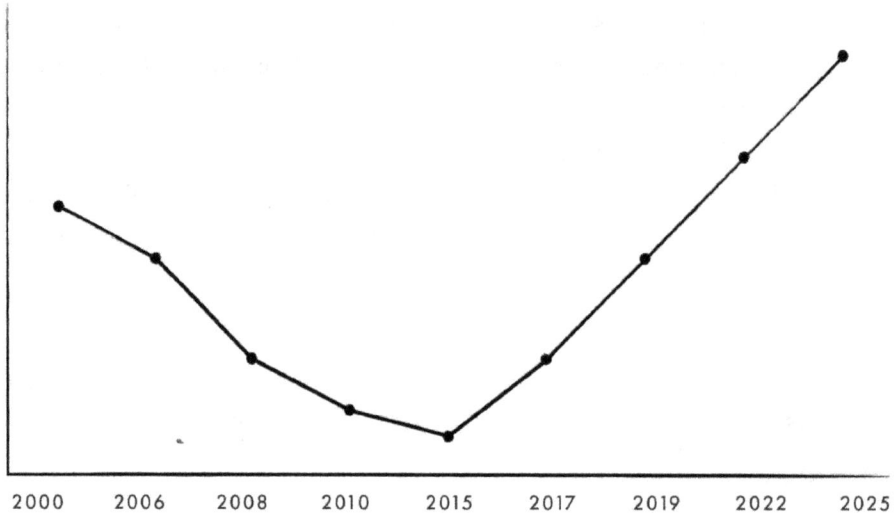

FURTILUX SHARE PRICE

Key Takeaways

- Diversity in leadership can help reenergize a failing business.
- Setting clear purpose and values provides crucial direction that can inspire business growth.
- LCAs and materiality assessments reveal the basis for change in the direction of greater sustainability.
- Sustainability software is critical for tracking progress on KPIs.
- Partnerships encourage innovation and bring reputational benefits.
- Becoming a Force for Good is good for business!

AN ACTION PLAN

FOR POSITIVE CHANGE

"The best way to predict the future is to invent it."
—Alan Kay, computer scientist at Xerox PARC

While it's true that bad news sells better than good, it's also true that the media have plenty of material to work with. Flip through the pages of almost any newspaper and it soon becomes clear that the world is facing some serious crises. Beyond the geopolitical worries, there are challenges of a social and environmental nature in just about every country across the globe.

The Stage Is Set for Business

I believe business can provide the spark and the machinery for change. Business can rise to the moment by meeting the real social and environmental challenges of the next decade and beyond. Business innovation—both in what we produce and how we act—can make the world a better place to live for all. That's not an idealistic fantasy. It's a matter of good business sense.

If the world is at a crossroads, then business can set the right direction to follow. Governments and consumers can also make a difference, but only business has the power to catalyze action efficiently and effectively at the scale and speed we need.

Throughout this book, I've celebrated companies that are already showing themselves as a Force for Good. They have found their Noble Purpose and put good intentions at the heart of their business strategy. They have recognized the shifting cultural sands and taken a lead in sustainability issues. And they're making use of data, expertise, and technological tools the likes of which we've never seen before.

Their companies are flourishing as a result. Consumers are buying into their Noble Purpose. People want to work for them. Investors are rewarding their foresight. They are moving ahead of regulation. I believe that these Lions of industry will be the ones that survive and grow into the future. They will gain more than their share of investment capital, business momentum, and talent, while the Ostriches that can't or won't adapt will find themselves cold-shouldered. Those that cover their eyes and ears will fall too far behind. They will be picked off or starved of resources.

For CEOs that aspire to be Lions and avoid the fate of the Ostriches, now is the time to analyze the Cycle of Good for signs of where it's heading. By identifying the sustainability issues that your business can materially help with, you can devise the innovations that will propel your business forward. With so many challenges, there are plenty of opportunities to do good.

Next Stop: Integrated Sustainability

The Holy Grail for sustainability is *integrated reporting*, whereby sustainability metrics are fully blended with financial metrics. It's a major undertaking, which requires mastery of data and also total buy-in from the CFO. To successfully achieve integrated reporting, your business needs to figure out which sustainability factors, along the whole supply chain, are material. You need to measure them, track improvement, engage with suppliers, and then deliver those results to the CFO.

Sustainability is just like any other investment in that you aim to net a return. For me, that's the major tipping point for sustainability. Financial directors increasingly understand that sustainability is not a cost. It's a revenue center and a cost-cutting center in the short and medium term. And in the long term, it's a crucial path to business success.

More and more business leaders agree. A recent survey by Gartner found that "86% of business leaders see sustainability as an investment that protects their organization from disruption. Additionally, 83% said sustainability program activities directly created both short- and long-term value for their organization, and 80% indicated that sustainability helped their organization optimize and reduce costs."[1] These figures show how far sustainability has advanced in recent years.

Increasingly, the fortunes of companies—and also the personal career trajectories of the executive team—will reflect their ability to hit targets on emissions, plastics, diversity, labor, and other sustainability metrics. As regulatory requirements increase and investors demand sustainability disclosures in exchange for access to capital, the reliance on enterprise sustainability management software and expertise that can automate and analyze data collection will increase, too.

In just a few years, and certainly by the end of this decade, the risk of supply chain disruption, consumer demands, and employee aspirations will turn integrated sustainability into business as usual. Board-level commitment to sustainability is rapidly increasing, with new sustainability teams led by CSOs. Major firms, from manufacturing to financial services and technology to life sciences, are turning to enterprise sustainability management platforms to provide a single source of truth that will make reporting auditable, traceable, and measurable. The changing nature of sustainability demands that these platforms are adaptable enough to accommodate evolving frameworks, disclosures, and regulations, with access to a comprehensive library of emission factors and LCA data. Any solution must navigate this complexity seamlessly.

Accelerate the Journey

As a final token, here is a seven-point action plan that I believe will help any business be a Force for Good.

1. Think from a cultural perspective, focusing not just on the business or even the customer point of view. How do you see sentiments changing in society? What does society expect of your company?
2. Envisage your company's environmental footprint through the eyes of the next generation who will be buying your products and services in 10 to 15 years.
3. Commission a full cradle-to-grave LCA of your most important products and services. Understand the full journey of your supply chain.
4. Conduct a materiality assessment to understand your stakeholders' priorities. You may find they differ from your own and also reveal opportunities for innovation and partnership.
5. Embrace enterprise sustainability management as a platform to operationalize change rather than an inconvenience or risk management exercise.
6. Break down the opportunities for innovation in sustainability across data, technology, and expertise. Try instigating at least one change across each of the three in your business.
7. Take the next step. Get busy with the stuff you can do right now. Don't let perfection get in the way of being proactive.

The Brightest Future Imaginable

The cultural wheel keeps turning. In the near future, some new technology may force every company to pivot in an unexpected direction. Or perhaps humankind will have united around a new idea. But at every turn, I am confident that entrepreneurs will have their hands on the wheel.

The newspapers often quote members of the public who are pessimistic about the future. They are troubled about the kind of world their descendants will inherit. I am not one of those people. I am confident that my own grand-children will live in a more sustainable world. I believe their generation will thrive in a society with opportunity for all. Business will have led the way. For them, sustainability and responsible investment will be business as usual.

For now, being a Force for Good means doing the right thing. The focus on sustainability is the next step in the journey for businesses to be centered around their customers, employees, communities, and investors. The rewards are ripe for the taking—not just for businesses, but also for the people and environment they influence. That's why I can't help but feel excited for a bright future. With business as a Force for Good, we can make it the brightest future imaginable.

BACKGROUND

ON CONTRIBUTORS

David Batchelor, Chairman of the Board at Sphera

D avid Batchelor has extensive international executive experience across his 40-year career in risk, insurance, and capital markets. He previously spent nearly two decades with the global risk advisory firm Marsh, including serving as the organization's vice chairman. In addition to his role as Sphera's non-executive chairman, David currently serves as non-executive chairman with Talbot Underwriting at Lloyd's of London. He is also an advisor to M&A and VC investing.

David has specific expertise in ESG, sustainability, and climate risk management and advancing diversity, equity, and inclusion programs. He is committed to raising transparency through board engagement on sustainability and is an alumnus of the UC Berkeley program ESG: Navigating the Board's Role. At Marsh, David led the group on emerging sustainability risks associated with climate change. He previously mentored female entrepreneurs through the Cherie Blair Foundation for Women in Emerging Markets and sponsored the Asia Resource Group at Marsh.

Deborah Cloutier, Chief Sustainability Officer at Legence

D eborah Cloutier, CRE®, the Chief Sustainability Officer at Legence, transforms buildings to pave the way for a sustainable future—faster. As the largest integrated provider of energy efficiency and sustainability solutions for the built environment, Legence represents over $2 billion in annual revenue and over 5,000 employees who specialize in helping clients decrease operating costs, reduce carbon emissions, and enhance occupancy well-being and productivity.

Deb is also the President and Founder of RE Tech Advisors, a Legence portfolio company, which designs and implements award-winning sustainability strategies, decarbonization plans and energy engineering solutions that improve asset and fund performance and comply with the evolving landscape of regulation. Under her leadership, RE Tech provides professional services to industry leaders with over $1.5 trillion of assets under management, representing 5.5 billion square feet of commercial properties. Deb is honored to be a Blackstone Senior Advisor and the recipient of *Entrepreneur Magazine*'s 100 Women of Influence, recognizing her as a trailblazer within the sustainability industry.

Dr. Michael Faltenbacher, Consulting Director, Transport and Mobility at Sphera

D r. Michael Faltenbacher joined Sphera in 2006 and is responsible for the transport and mobility sector as consulting director. With more than 20 years of experience in the field of life cycle analysis of transport and energy systems, he has extensive knowledge on the automotive sector, covering the transport of passengers and goods using conventional and alternative fuel and drive systems. His work focuses on the technical, ecological, and economic comparison of fuel and propulsion technologies for light and heavy-duty vehicles.

Michael holds a master's degree in chemical engineering (Dipl. Ing.) from the University of Stuttgart and received his doctorate in mechanical engineering from the University of Stuttgart on the topic of life cycle assessment of hydrogen-powered fuel cell buses in comparison to diesel and CNG buses. He has authored and coauthored various white papers and publications on the introduction of zero-emission vehicles in light- and heavy-duty vehicle fleets, including buses for public transport and trucks for long-haul goods transport.

Alex Gershenson, Global Supply Chain Sustainability Advisor at Sphera

Alex Gershenson received his doctorate in environmental studies from UC Santa Cruz, with a focus on climate science and policy. He is a well-known expert in responsible sourcing, sustainability, carbon accounting, environmental policy, and economics. Alex is a published author and a frequent speaker at conferences that focus on responsible sourcing and supply chain management. Alex is the founding CEO of SupplyShift, which was acquired by Sphera. Prior to starting SupplyShift, Alex was a professor of policy and science at San Jose State University and a co-founder of an environmental consulting firm, EcoShift Consulting, which focused on climate and carbon consulting for enterprise and government clients.

Philippe Guillard, Vice President, Global Solutions and Operations at Sphera

P hilippe Guillard has been with Sphera for nearly 20 years. His career has spanned operational risk management, environmental sustainability management, enterprise software management, industrial automation, and control and process system integration. He has worked with companies in over 27 countries, aiding their mission toward enabling operational sustainability and operational risk management.

Philippe holds a B.Eng. degree in chemical engineering from McGill University (Montreal, Canada) and is a registered member of the Professional Engineers of Ontario (PEO).

Gunjan Khera, Global Head of Platform Engineering, Analytics and AI at Sphera

G unjan Khera leads the SpheraCloud platform's strategy and execution in concert with clients' sustainability goals and product road maps. Gunjan has a unique blend of experience and expertise in enterprise-grade technology, product management, and analytics across large corporations, which allows him to bring modern platform thinking to Sphera.

Gunjan graduated from the University of Illinois Chicago with a bachelor's degree in computer science. Over the course of his career, he has been privileged to have worked with some great people and companies, including Accenture, Marriott, Walgreens, McDonald's, Nielsen. and many more. Gunjan enjoys semipro gourmet cooking, photography, and engaging in spirited debate around entrepreneurship, circle-of-life, and technology advancements for humankind.

Stefan Kupferschmid, Senior Sustainability Consultant—Sector Lead, Automotive at Sphera

S tefan Kupferschmid is a senior sustainability consultant and mechanical engineer with more than eight years of professional experience. He joined Sphera in 2017 from the R&D department of a tier-1 automotive supplier. At Sphera, he has been working on multiple projects in the field of sustainable fuels and mobility. Since 2020, he has been responsible for sustainability consulting projects in the automotive sector with a strong focus on product sustainability and life cycle assessment (LCA).

Stefan has been involved in a multitude of LCA projects with industrial clients and international organizations on the environmental assessment of entire cars as well as automotive components. As sector lead for automotive, he has a strong understanding of the most pressing topics the industry is facing when it comes to sustainability. His main areas of focus are the development of decarbonization strategies and enabling clients to fulfill OEM requirements for the reporting of product carbon footprints.

Elizabeth Lewis, Managing Director and Head of ESG for Blackstone Infrastructure Partners

P rior to joining Blackstone, Elizabeth Lewis was at the International Finance Corporation (IFC), the private sector part of the World Bank Group, leading engagement with investors, NGOs, governments, and other stakeholders on climate change and diversity. Prior to joining the IFC, Elizabeth was a partner and director of strategy and business development for Terra Alpha Investments, having previously established the World Resources Institute's (WRI's) Sustainable Investing Program and served as the WRI's head of sustainable investing. Earlier in her career, Elizabeth was a principal at the Global Environment Fund, a private equity fund focused on clean energy and sustainable forestry. She started her career at Booz Allen Hamilton, advising clients in their evaluation of alternative energy technologies and solutions.

Elizabeth holds an A.B. in environmental science and public policy from Harvard College and an MBA from Harvard Business School. She serves as chair of the board of trustees for The Nature Conservancy Maryland/D.C. She also serves as a member of the Harvard Business School Alumni Board and the Harvard Alumni Association Schools and Scholarships Committee and is the class chair of the John Harvard Society.

Hari Osofsky, Dean and Myra and James Bradwell Professor of Law at Northwestern Pritzker School of Law and Professor of Environmental Policy and Culture (courtesy) at the Weinberg College of Arts and Sciences at Northwestern University

Dean Osofsky received her B.A. from Yale College, a J.D. from Yale Law School, and a Ph.D. in geography from the University of Oregon. Prior to joining Northwestern University, Hari served as dean of Penn State Law and the Penn State School of International Affairs and on the faculties of the University of Minnesota Law School, Washington and Lee University School of Law, the University of Oregon School of Law, and Whittier Law School. She clerked for Judge Dorothy W. Nelson of the Ninth Circuit Court of Appeals.

As dean, she is deeply committed to building legal education for a changing society through inclusive collaboration and holds leadership roles in numerous professional organizations. Hari's 50+ publications focus on improving governance and addressing injustice in energy and climate change regulation.

Amisha Parekh, Managing Director and Global Head of ESG for Private Equity at Blackstone

A misha Parekh leads ESG diligence, policy development, and strategy and reporting for Blackstone's Private Equity Group.

Amisha joined Blackstone from Bloomberg LP, where she led ESG data acquisition and curation efforts. Prior to Bloomberg, she was a senior manager at Deloitte. Amisha is the coauthor of *High Performance Hospitality: Sustainable Hotel Case Studies* (Educational Institute, 2013) and was an adjunct professor of business strategy for sustainability at Glasgow Caledonian New York College.

Dane Parker, Board Member at Sphera

A s the former chief sustainability officer at General Motors Co. (GM), Dane Parker was the first chief sustainability officer at one of the largest auto manufacturers in the world and was instrumental in creating its goals for its sustainability journey.

Dane has extensive experience in sustainability and setting environmental and emissions targets. He was a leading force behind GM's plan to become carbon neutral by 2040 and its aspiration to have zero emissions from all new light-duty vehicles by 2035. Prior to GM, Dane was vice president, global environment, health and safety (EHS), real estate and facilities, for Dell Inc., where he established the global EHS team and achieved world-class results in both safety and environmental performance.

Preyasi Patel, Director, Client Services—Corporate Sustainability EMEA at Sphera

P reyasi Patel, director of client services for corporate sustainability at Sphera, has 10+ years of experience in the sustainability space, working with large multinational organizations to support their increasingly complex and detailed reporting requirements through a combination of consulting and software. She works with organizations that are in varying phases of their sustainability journeys to optimize data-gathering, align with changing regulatory demands, and drive value from information to make critical business decisions relating to sustainability. In addition, she is responsible for advancing Sphera's internal ESG program.

Jacqueline Peel, Professor at Melbourne Law School and Director, Melbourne Climate Futures at University of Melbourne

P rofessor Jacqueline Peel is a leading, internationally recognized expert in environmental and climate change law. Her scholarship encompasses international, transnational and national dimensions, as well as interdisciplinary aspects of the law/science relationship in the environmental field and risk regulation. She is currently the director of Melbourne Climate Futures, the University of Melbourne's climate change initiative.

The author and co-author of several books and numerous articles, Jacqueline has also been a visiting scholar at the Berkeley Law School's Center for Law, Energy and Environment and also at Stanford Water in the West, Stanford University. Together with Dean Hari Osofsky, Jacqueline provides evaluation and research consultancy services to the U.K.-based Children's Investment Fund Foundation for its grants on strategic climate change litigation. She has also carried out consultancies on environmental law and policy issues for organizations such as UNEP, ClientEarth, and the Secretariat of the Pacific Regional Environmental Programme.

Jacqueline has been an active contributor to national and international policy processes, most recently as a lead author on Working Group III (mitigation) of the Intergovernmental Panel on Climate Change. She was elected a fellow of the Academy of the Social Sciences in Australia in 2019 and has received several prestigious awards such as a Fulbright Scholarship, NYU Hauser Scholarship and the Morrison Prize 2018 for her award-winning article with Dean Osofsky on Energy Partisanship.

Stefan Premer, Principal Consultant and Global Lead of Climate Strategy at Sphera

Stefan Premer is a principal consultant for Sphera's sustainability consulting organization and Sphera's domain lead for climate strategy and net zero. He develops sustainability and corporate climate strategies, science-based climate, and net-zero targets and supply chain decarbonization initiatives. Stefan has expertise in sustainability regulatory developments and reporting, corporate carbon accounting, Scope 3 method development, decarbonization studies, climate risk, and scenario modeling, and also provides insights on value chain mitigation schemes. He has managed various strategic, large-scale projects across sectors.

Stefan graduated with his dissertation on business model innovation in clean technology initiatives in developing countries and holds an M.Sc. in business administration: strategy and management in international organizations from Linköping University in Sweden. Furthermore, he has project experience from several sustainability NGOs in South Asia, Latin America, and Africa.

Heiko Schwarz, Global Supply Chain Risk Advisor at Sphera

Heiko Schwarz has worked in the software market for 20 years, in various areas of strategic purchasing and supply chain.

With his expertise in procurement, supply management, risk management, and digital transformation, Heiko has successfully helped supply chain and procurement organizations implement solutions to increase performance, reduce costs, and minimize risks. In his position as sales director at IBM/Emptoris, he was responsible for global sales, thought leadership for strategic procurement, and transfer of know-how for the Supplier Management product line.

In 2013, he cofounded riskmethods, to focus on delivering an innovative software-as-a-service solution for comprehensive supply chain risk management. riskmethods was acquired by Sphera in October 2022.

Dr. Rajesh Singh, Managing Director, India and Southeast Asia at Sphera

Dr. Rajesh Singh works in the capacity of managing director (business development and services), India, Asia Pacific, and the Middle East at Sphera. He has more than 30 years of experience in the areas of sustainability strategy, life cycle assessment (LCA), eco-design, climate change and decarbonization strategy, and was a founding member of thinkstep India, which was acquired by Sphera.

Rajesh is a civil engineer, postgraduate environmental engineer from the Indian Institute of Technology Kharagpur, and has a doctorate in sustainability management from the Indian Institute of Technology Bombay. He has also published more than 25 papers in international journals and presented at multiple national and international conferences.

Jürgen Stichling, Director, Solution Consulting— Product Sustainability at Sphera

J ürgen Stichling supports Sphera clients and prospects in developing the best solution for successfully achieving their sustainability targets.

With over 30 years of experience in life cycle assessment (LCA) and sustainability, Jürgen is a well-known global industry expert for LCA with a focus on automotive, transport and logistics, lightweight materials, and manufacturing of complex products. As a consultant, he has worked on LCA projects that range from strategic material and process selection to benchmarking studies for clients in both Europe and Asia. For more than 25 years, Jürgen has supported global automotive manufacturers, rail companies, tier-1 suppliers and the aerospace industry in implementing life cycle thinking in their business processes.

Jürgen holds a Dipl.-Ing. (Master) degree in chemical engineering. He studied at the University of Stuttgart, Germany.

Candace Vogler, David B. and Clara E. Stern Professor of Philosophy and Professor in the College, University of Chicago

P rofessor Vogler is the principal investigator on Virtue, Happiness, and the Meaning of Life, a project funded by the John Templeton Foundation. She has authored two books—*John Stuart Mill's Deliberative Landscape: An Essay in Moral Psychology* (Routledge, 2001) and *Reasonably Vicious* (Harvard University Press, 2002)—and essays in ethics, social and political philosophy, philosophy and literature, cinema, psychoanalysis, gender studies, sexuality studies, and other areas. Her research interests are in practical philosophy (particularly the moral philosophy indebted to Elizabeth Anscombe), practical reason, Kant's ethics, Marx, and neo-Aristotelian naturalism.

Candace received a B.A. from Mills College and a Ph.D. from the University of Pittsburgh.

SOURCE NOTES

1. Introducing Sustainability as a Force for Good

[1] Patagonia, "Company History," accessed December 4, 2024. eu.patagonia.com/gb/en/company-history/

[2] "Yvon Chouinard: Let My People Go Surfing" (2013), UCLA Environment, youtube.com/watch?v=EHS2X-KoN_w

[3] "Interview with Yvon Chouinard, 'Founding Patagonia & Living Simply" (2016), Commonwealth Club World Affairs. youtube.com/watch?v=ZQlu95rzUTM.

[4] "Yvon Chouinard: Let My People Go Surfing" (2013), UCLA Environment, youtube.com/watch?v=EHS2X-KoN_w

[5] "The New Footprint Chronicles - Patagonia's supply chain examined" (2015), Patagonia, youtube.com/watch?v=JIC9DUkbic8&t=24s

[6] Vincent Stanley and Yvon Chouinard, "The Future of the Responsible Company – What we've learnt from Patagonia's first 50 years" (Patagonia Works, 2023).

[7] Patagonia, "B Lab," accessed December 4, 2024. https://eu.patagonia.com/gb/en/b-lab.html?srsltid=AfmBOoqIIsuOVRl-UL6jSHk5dKBhhcxmdmNXbN863PuIuASarjA-TSAY

[8] UNEP, "US outdoor clothing brand Patagonia wins UN Champions of the Earth award," September 24, 2019. https://www.unep.org/news-and-stories/press-release/us-outdoor-clothing-brand-patagonia-wins-un-champions-earth-award

[9] Stanley and Chouinard, *op. cit.*

[10] Patagonia, "Patagonia's next chapter: Earth is now our only shareholder," September 14, 2022. patagoniaworks.com/press/2022/9/14/patagonias-next-chapter-earth-is-now-our-only-shareholder

[11] Stanley and Chouinard, *op. cit.*

[12] Jim Collins, *Good to Great: Why Some Companies Make the Leap… and Others Don't.* (New York, NY: Harper Business, 2001).

[13] Klaus Schwab and Peter Vanham, "What is stakeholder capitalism?" *World Economic Forum,* January 22, 2021. https://www.weforum.org/agenda/2021/01/klaus-schwab-on-what-is-stakeholder-capitalism-history-relevance

[14] The National Intelligence Council, "Global Trends 2040: A More Contested World," March 2021. https://www.dni.gov/files/ODNI/documents/assessments/GlobalTrends_2040.pdf

[15] Chris Zook and James Allen, *The Founders Mentality.* (Boston, Mass.: Harvard Business Review Press, 2016).

[16] Martin Reeves, Simon Levin and Daichi Ueda, "The Biology of Corporate Survival," *Harvard Business Review* (January–February 2016): 46–55. https://hbr.org/2016/01/the-biology-of-corporate-survival

[17] *21st Annual Edelman Trust Barometer.* Edelman (January 2021). https://www.edelman.com/trust/2021-trust-barometer

[18] Raphael Bemporad et al., *Radically Better Future: The Next Gen Reckoning Report* (December 2020) https://globescan.wpenginepowered.com/wp-content/uploads/2020/12/BBMG_GlobeScan_Radically-Better-Future-Report_2020-1.pdf

[19] Aaron Hurst et al., *Purpose at Work, 2016 Global Report.* LinkedIn and Imperative, accessed April 2023. https://business.linkedin.com/content/dam/me/business/en-us/talent-solutions/resources/pdfs/purpose-at-work-global-report.pdf.

[20] "The Power of Strategic Purpose," strategy&, accessed April 2023, https://www.strategyand.pwc.com/gx/en/unique-solutions/capabilities-driven-strategy/approach/research-motivation.html

21 "Undivided: The Porter Novelli/Cone Gen Z Purpose Study, 2019" Porter Novelli, accessed April 2023. https://engageforgood.com/portern-novelli-cone-genz-2019/

22 "The Business Case for Purpose," *Harvard Business Review* (October 2016). https://assets.ey.com/content/dam/ey-sites/ey-com/en_gl/topics/digital/ey-the-business-case-for-purpose.pdf

23 "Putting Purpose to Work: A study of purpose in the workplace" PwC (June 2016) https://www.pwc.com/us/en/purpose-workplace-study.html

24 George Serafeim and Claudine Gartenberg "The Type of Purpose That Makes Companies More Profitable" *Harvard Business Review* (October 21, 2016) https://hbr.org/2016/10/the-type-of-purpose-that-makes-companies-more-profitable

25 Claudine Madras Gartenberg, Andrea Prat and George Serafeim "Corporate Purpose and Financial Performance" *Organization Science* (October 9, 2018), 30(1), pp.1-18. https://ssrn.com/abstract=2840005

26 Dominic Barton et al., "Measuring the Economic Impact of Short-Termism" McKinsey & Co. February 8, 2017. https://www.mckinsey.com/featured-insights/long-term-capitalism/where-companies-with-a-long-term-view-outperform-their-peers

27 "Regulation database," Principles for Responsible Investment. Accessed March 2024. https://www.unpri.org/policy/global-policy/regulation-database

28 Margarita Pirovska and Fiona Stewart, "Supporting policy makers and regulators to build a sustainable financial system," PRI blog, December 2020. https://www.unpri.org/pri-blog/supporting-policy-makers-and-regulators-to-build-a-sustainable-financial-system/6930.article

29 "Global Sustainable Investment Review," Global Sustainable Investment Alliance, November 2023. https://www.gsi-alliance.org/members-resources/gsir2022/

30 Sphera Sustainability Survey, September 2021. https://sphera.com/sustainability/sustainability-survey-2021/#report

31 Reuters, "Data centers could use 9% of US electricity by 2030, research institute says," May 29, 2024. https://www.reuters.com/business/energy/data-centers-could-use-9-us-electricity-by-2030-research-institute-says-2024-05-29/

2. The Cycle of Good

[1] *Oxford Bible Commentary,* ed. John Barton and John Muddiman*, (Oxford:* Oxford University Press, 2001) Current online edition, 2022. https://www.oxfordreference.com/display/10.1093/acref/9780198755005.001.0001/acref-9780198755005;jsessionid=00FFFAC04AD631385591833822065999

[2] Kern, *The Culture of Time and Space, 1880–1918* (Boston, Mass: Harvard University Press, 2003)

[3] Kern, *The Culture of Time and Space*

[4] "U.S. Census Bureau History: Alexander Graham Bell," U.S. Census Bureau, March 2021. https://www.census.gov/history/www/homepage_archive/2021/march_2021.html

[5] The Editors of Encyclopaedia Britannica, "Industrial Revolution," *Encyclopaedia Britannica*, last updated April 4, 2024. https://www.britannica.com/event/Industrial-Revolution#ref347982

[6] Lizabeth A Cohen, "A Consumer's Republic: The Politics of Mass Consumption in Post-war America," *Journal of Consumer Research* 31(1): 236-239 (2004). https://dash.harvard.edu/bitstream/handle/1/4699747/cohen_conrepublic.pdf

[7] U.S Bureau of Labor Statistics. https://data.bls.gov/

[8] Stanley Lebergott, *Pursuing Happiness: American Consumers in the Twentieth Century.* (Princeton, NJ: Princeton University Press, 1993)

[9] History.com Editors, "G.I. Bill," *History.com,* last revised June 7, 2019. https://www.history.com/topics/world-war-ii/gi-bill

[10] Claire Boyte-White, "The Basic Economic Effects World War II Had on the Global Economy," *Investopedia.com*, last revised June 12, 2023. https://www.investopedia.com/ask/answers/112814/how-did-world-war-ii-impact-european-gdp.asp

[11] "Best Global Brands 2023," Interbrand, accessed March 2024. https://interbrand.com/best-global-brands/

[12] Vicki Broadbent, "How IBM misjudged the PC revolution," *BBC News,* last revised April 4, 2005. http://news.bbc.co.uk/1/hi/business/4336253.stm

[13] Lou Gerstner, "Lou Gerstner on corporate reinvention and values," interview by **Ian Davis** and **Tim Dickson**. *McKinsey Quarterly*, September 1, 2014 https://www.mckinsey.com/featured-insights/leadership/lou-gerstner-on-corporate-reinvention-and-values#/

[14] Entrepreneur staff, "Charles Lazarus: Toy Titan," October 10, 2008. https://www.entrepreneur.com/leadership/charles-lazarus/197660

[15] Erin Blakemore, "Inside the Rise and Fall of Toys "R"' Us," *History.com*, last revised May 10, 2023. https://www.history.com/news/toys-r-us-closing-legacy

[16] Michael S Rosenwald, "Toys R U.S. founder Charles Lazarus has died. Here's how he built — and lost — an empire," *The Washington Post*, March 22, 2018. https://www.washingtonpost.com/news/retropolis/wp/2017/09/19/toys-r-us-the-birth-and-bust-of-a-retail-empire/

[17] Rosenwald, "Toys R U.S. founder"

[18] "Toys R U.S. … not dead yet," The Associated Press. October 3, 2018. https://apnews.com/article/toys-business-us-news-north-america-b66b906e9c2a40bfa9939828e3366b95

[19] PR Newswire, "WHP Global Announces Toys"R"Us® Retail Expansion to Air, Land and Sea", September 29, 2023. https://www.prnewswire.com/news-releases/whp-global-announces-toysrus-retail-expansion-to-air-land-and-sea-301943033.html

[20] Jennifer Tonti, "Americans Want to See Greater Transparency on ESG Issues and Support Federal Requirements for Increasing Disclosure," Just Capital, February 2022. https://com-justcapital-web-v2.s3.amazonaws.com/pdf/JUSTCapital_CorporateDisclosureStandardsSurveyReport_SSRS_Ceres_PublicCitizen_Feb2022.pdf

3. The Formula for Good

[1] "The true death toll of COVID-19," World Health Organization, accessed April 2023, https://www.who.int/data/stories/the-true-death-toll-of-covid-19-estimating-global-excess-mortality

[2] Andrea Shalal, "IMF sees cost of COVID pandemic rising beyond $12.5 trillion estimate," Reuters, January 20, 2022, https://www.reuters.com/business/imf-sees-cost-covid-pandemic-rising-beyond-125-trillion-estimate-2022-01-20/

[3] Dave Roos, "How a New Vaccine Was Developed in Record Time in the 1960s," *History.com*, last updated October 4, 2023. https://www.history.com/news/mumps-vaccine-world-war-ii

[4] "Understanding How COVID-19 Vaccines Work," Centres for Disease Control and Prevention, last revision September 22, 2023, https://www.cdc.gov/coronavirus/2019-ncov/vaccines/different-vaccines/how-they-work.html

[5] David Cox, "How mRNA went from a scientific backwater to a pandemic crusher," *Wired.com*, December 2, 2020, https://www.wired.co.uk/article/mrna-coronavirus-vaccine-pfizer-biontech

[6] "Pushing boundaries to deliver COVID-19 vaccine across the Globe," Astra Zeneca, accessed April 2023, https://www.astrazeneca.com/what-science-can-do/topics/technologies/pushing-boundaries-to-deliver-covid-19-vaccine-accross-the-globe.html

[7] "Using Data Analytics to Accelerate COVID-19 Vaccine Development," Sartorius, June 18, 2020, https://www.sartorius.com/en/knowledge/science-snippets/using-data-analytics-to-accelerate-covid-19-vaccine-development-549432

[8] Josh Holder, "Tracking Coronavirus Vaccinations Around the World," *New York Times*, last revision March 13, 2023. https://www.nytimes.com/interactive/2021/world/covid-vaccinations-tracker.html

[9] Oliver J Watson et al., "Global impact of the first year of COVID-19 vaccination: a mathematical modelling study," *The Lancet*, Volume 22, Issue 9, P1293-1302, September 2022. Published: June 23, 2022. https://www.thelancet.com/journals/laninf/article/PIIS1473-3099(22)00320-6/fulltext#%20

[10] Cem Çakmaklı et al., "The Economic Case for Global Vaccinations," International Chamber of Commerce, January 25, 2021 https://iccwbo.org/publication/the-economic-case-for-global-vaccinations/

[11] "How unlocking health data shaped the COVID-19 vaccine rollout," HDRUK, February 16, 2023. https://www.hdruk.ac.uk/case-studies/how-unlocking-health-data-shaped-the-covid-19-vaccine-rollout-2/

12 Elaine Robertson et al, "Predictors of COVID-19 vaccine hesitancy in the UK household longitudinal study," *National Library of Medicine*, May 2021, https://www.ncbi.nlm.nih.gov/pmc/articles/PMC7946541/

13 Shadaab K et al., "Technology and Sustainability Market. Global Opportunity Analysis and Industry Forecast 2021-2030," Allied Market Research, October 2021. https://www.alliedmarketresearch.com/green-technology-and-sustainability-market-A06033

14 PW Kingsford, "James Watt." *Encyclopaedia Britannica*, January 15, 2023. https://www.britannica.com/biography/James-Watt.

15 Rochelle Forrester, "The Invention of the Steam Engine." *SocArXiv*. October 10, 2019, doi:10.31235/osf.io/fvs74.

16 Chloe Sorvino, "The Gilded Age Family That Gave It All Away: The Carnegies," *Forbes*, July 8, 2014. https://www.forbes.com/sites/chloesorvino/2014/07/08/whats-become-of-them-the-carnegie-family/?sh=21585ace7b55

17 Editors of Encyclopaedia, "Andrew Carnegie." *Encyclopaedia Britannica*, January 6, 2023 https://www.britannica.com/biography/Andrew-Carnegie

18 Editors of Encyclopaedia, "Andrew Carnegie." *Encyclopaedia Britannica*, January 6, 2023 https://www.britannica.com/biography/Andrew-Carnegie

19 "Carnegie on how to get rich," St. Louis Globe-Democrat, October 15, 1899. https://www.newspapers.com/article/st-louis-globe-democrat-carnegieadvice/84452036/

20 Bengt-Arne Vedin, "The transistor – an invention ahead of its time," Telefonaktiebolaget LM Ericsson and Centre for Business History, accessed April 2023, https://www.ericsson.com/en/about-us/history/products/other-products/the-transistor--an-invention-ahead-of-its-time

21 "Claude Shannon: The Father of Information Theory," History of Data Science, Dataiku, May 6, 2021, https://www.historyofdatascience.com/claude-shannon/

22 David Tse, "How Claude Shannon Invented the Future," *Quanta Magazine, last revised January 4, 2021,* https://www.quantamagazine.org/how-claude-shannons-information-theory-invented-the-future-20201222/

23 "Claude Shannon: The Father of Information Theory," History of Data Science.

24 Robert Price, "A conversation with Claude Shannon: One man's approach to problem solving," IEEE Communications Magazine 22 (1984): p. 126.

[25] Zia Khan, "Motivating Change: How the Data Revolution Can Feed the Next Green Revolution," Rockefeller Foundation, August 29, 2018, https://www.rockefeller-foundation.org/blog/motivating-change-data-revolution-can-feed-next-green-revolution/

[26] Gail MacLeitch, "Video conferencing statistics: usage and trends 2022," *Quickblox*, 30 September, 2022. https://quickblox.com/blog/video-conferencing-statistics-usage-and-trends/

[27] Paul Daugherty et al., "Make the leap, take the lead. Tech strategies for innovation and growth," Accenture Research, accessed April 2023, https://www.accenture.com/us-en/insights/technology/scaling-enterprise-digital-transformation?c=acn_glb_futuresystemsmediarelations_12144611&n=mrl_0421

[28] MIT Technology Review, "New Approaches to the tech talent shortage," September 21, 2023. https://www.technologyreview.com/2023/09/21/1079695/new-approaches-to-the-tech-talent-shortage/

[29] "Global Tech Report 2022," KPMG International, September 2022. https://home.kpmg/xx/en/home/insights/2022/09/kpmg-global-tech-report-2022.html

[30] "Management Consulting Services Global Market Report 2022," The Business Research Company, December 2021, https://finance.yahoo.com/news/management-consulting-services-global-market-121800506.html

[31] Sari Pekkala Kerr et al., "Global Talent Flows," *Journal of Economic Perspectives*, 30 (4): 83-106, 2016, accessed April 2023 https://www.aeaweb.org/articles?id=10.1257/jep.30.4.83

[32] Ethan Baron, "H-1B: Foreign citizens make up nearly three-quarters of Silicon Valley tech workforce, report says," Mercury News, January 17, 2018, last revised May 8, 2018, https://www.mercurynews.com/2018/01/17/h-1b-foreign-citizens-make-up-nearly-three-quarters-of-silicon-valley-tech-workforce-report-says/

[33] Rocio Lorenzo and Martin Reeves, "How and Where Diversity Drives Financial Performance," *Harvard Business Review*, January 30, 2018. https://hbr.org/2018/01/how-and-where-diversity-drives-financial-performance

[34] Ejaz Ghani, "The global talent race heats up as countries and businesses compete for the best and brightest," World Economic Forum, November 23, 2018, https://www.weforum.org/agenda/2018/11/the-global-talent-race/

[35] Edward Tufte, *The Visual Display of Quantitative Information* (Graphics Press, 2nd Edition 2001)

[36] H.P. Luhn, "A business intelligence system," IBM Journal of Research and Development, Volume 2, Issue 4, pages 314-319, October 1958.

[37] "Volume of data/information created, captured, copied, and consumed worldwide from 2010 to 2020, with forecasts from 2021 to 2025" June 2021, https://www.statista.com/statistics/871513/worldwide-data-created/

[38] Bernard Marr, "The 5 Biggest Data Science Trends In 2022," *Forbes.com*, October 4, 2021, https://www.forbes.com/sites/bernardmarr/2021/10/04/the-5-biggest-data-science-trends-in-2022/?sh=702c63d740d3

[39] "The data-driven enterprise of 2025," McKinsey and Co., January 28, 2022, https://www.mckinsey.com/capabilities/quantumblack/our-insights/the-data-driven-enterprise-of-2025

4. Mobility for Good

[1] Christian Wolmar, *The Great Railroad Revolution: The History of Trains in America*, (PublicAffairs, September 2012)

[2] Gilbert King, "Where the Buffalo No Longer Roamed," *The Smithsonian Magazine*, July 17, 2012, https://www.smithsonianmag.com/history/where-the-buffalo-no-longer-roamed-3067904/

[3] "Transcontinental Railroad." *HISTORY.com,* last revised September 11, 2019, https://www.history.com/topics/inventions/transcontinental-railroad

[4] Patrick J Kiger, "*10 Ways the Transcontinental Railroad Changed America,*" HISTORY.com, last revised July 25, 2023, https://www.history.com/news/transcontinental-railroad-changed-america

[5] Weissenbacher, Manfred, *Sources of Power: How Energy Forges Human History.* (Praeger, 2009).

[6] "6 Key Differences Between American and European Rail Systems," Seminole Railways, accessed April 4, 2023, https://www.floridarail.com/news/6-key-differences-between-american-and-european-rail-systems/

[7] "The Japanese Maglev: World's fastest bullet train," *Japan Rail Pass,* November 22, 2014. https://www.jrailpass.com/blog/maglev-bullet-train

[8] Northeast Maglev, accessed March 2024. https://northeastmaglev.com

[9] Rail Technical Strategy "Desired outcomes." https://railtechnicalstrategy.co.uk/desired-outcomes/

[10] Hervé Macé, "Three major challenges for the railway market in 2021," Segula Technologies, 10 March 2021, https://www.segulatechnologies.com/en/news/three-major-challenges-for-the-railway-market-in-2021/

[11] "Poll results show Americans continue to strongly support rail development," Global Railway Review, 15 September 2020, https://www.globalrailwayreview.com/news/109026/poll-americans-support-rail-development/

[12] Henry Ford et al., *My Life and Work*, (Garden City, NY, Doubleday, Page & Company, 1922.) https://www.loc.gov/item/22026971/.

[13] History.com, "Henry Ford," last updated March 26, 2020. https://www.history.com/topics/inventions/henry-ford

[14] "Ford's assembly line starts rolling," *History.com*, accessed April 2024. https://www.history.com/this-day-in-history/fords-assembly-line-starts-rolling#

[15] "Clarence Avery," The Automotive Hall of Fame, accessed April 4, 2023, https://www.automotivehalloffame.org/honoree/clearence-w-avery/

[16] John Bell Rae and Alan K Binder, "Automotive industry," *Encyclopaedia Britannica*, last updated March 9, 2024. https://www.britannica.com/technology/automotive-industry

[17] David Pietrusza, "Henry Ford and Alfred P. Sloan: Industrialization and Competition," Bill of Rights Institute, accessed April 2024. https://billofrightsinstitute.org/essays/henry-ford-and-alfred-p-sloan-industrialization-and-competition#

[18] The Editors of Encyclopaedia Britannica, "Alfred P. Sloan, Jr." Encyclopaedia Britannica, last updated April 9, 2024. https://www.britannica.com/biography/Alfred-P-Sloan-Jr

[19] Steve Blank, "Apple's Marketing Playbook Was Written in the 1920s," *The Atlantic*, October 26, 2011. https://www.theatlantic.com/business/archive/2011/10/apples-marketing-playbook-was-written-in-the-1920s/247417/

Source Notes

20 Clarke V. Simmons "The Sloan Legacy," *London Business School*, December 1, 2009. https://www.london.edu/think/the-sloan-legacy

21 Clarke V. Simmons, "The Sloan Legacy."

22 Mary Bellis, "The history of crash test dummies," *ThoughtCo.*, last updated April 17, 2019. https://www.thoughtco.com/history-of-crash-test-dummies-1992406

23 Douglas Bell, "Volvo's Gift To The World, Modern Seat Belts Have Saved Millions Of Lives,*" Forbes.com*, August 13, 2019. https://www.forbes.com/sites/douglasbell/2019/08/13/60-years-of-seatbelts-volvos-great-gift-to-the-world/?sh=8bd9f8f22bcc

24 "Seat belts," NHTSA, accessed April 4, 2023, https://www.nhtsa.gov/risky-driving/seat-belts

25 Volvo Cars, "Volvo Cars to go all electric," July 5, 2017. https://www.media.volvocars.com/global/en-gb/media/pressreleases/210058/volvo-cars-to-go-all-electric#:~:text=Volvo%20Cars%2C%20the%20premium%20car,core%20of%20its%20future%20business

26 Marcus Lu, "The Decline of U.S. Car Production,*" Visual Capitalist*, December 3, 2021, https://www.visualcapitalist.com/the-decline-of-u-s-car-production/

27 Mathilde Carlier, "Estimated U.S. market share held by selected automotive manufacturers in 2023," *Statista*, March 11, 2024. https://www.statista.com/statistics/249375/us-market-share-of-selected-automobile-manufacturers/

28 "36 Automotive Industry Statistics [2023]: Average Employment, Sales, And More," *Zippia.com*, March 15, 2023, https://www.zippia.com/advice/automotive-industry-statistics/

29 Chuck Tannert, "John DeLorean Reinvented The Dream Car. Then He Totaled It," *Forbes.com*, July 26, 2019, https://www.forbes.com/wheels/news/john-delorean-reinvented-the-dream-car-then-he-totaled-it/

30 "Most valuable brands within the automotive sector worldwide as of 2022, by brand value," Statistica, June 2022, https://www.statista.com/statistics/267830/brand-values-of-the-top-10-most-valuable-car-brands/

31 Mike Musgrove, "An Electric Car with Juice," *Washington Post*, July 22, 2006. https://www.washingtonpost.com/wp-dyn/content/article/2006/07/21/AR2006072101515.html

[32] Jens Meiners, "2009 Tesla Roadster," *Car and Driver*, August 21, 2008. https://www.caranddriver.com/reviews/a15142083/2009-tesla-roadster-first-drive-review/

[33] Andrew Yeadon and Angus MacKenzie, "2013 Motor Trend Car of the Year: Tesla Model S," *Motortrend*, December 10, 2012. https://www.motortrend.com/news/2013-motor-trend-car-of-the-year-tesla-model-s/

[34] Elon Musk, "All our patent are belong to you," *Tesla*, June 12, 2014. https://www.tesla.com/en_gb/blog/all-our-patent-are-belong-you

[35] "Electric vehicle battery prices are expected to fall almost 50% by 2026," Goldman Sachs, October 7, 2024. https://www.goldmansachs.com/insights/articles/electric-vehicle-battery-prices-are-expected-to-fall-almost-50-percent-by-2025#

[36] International Energy Association, "Global EV Outlook 2023, Catching up with climate ambitions," https://iea.blob.core.windows.net/assets/dacf14d2-eabc-498a-8263-9f97fd5dc327/GEVO2023.pdf

[37] Kingsmill Bond et al, "X-Change: Cars. The end of the ICE age," *RMI*, September 2023. https://rmi.org/wp-content/uploads/dlm_uploads/2023/09/x_change_cars_report.pdf

[38] Joann Muller, "Tesla Inside: How Elon Musk could control more of the EV industry," *Axios*, June 23, 2023. https://www.axios.com/2023/06/23/tesla-elon-musk-ev-charging

[39] Alexandra Villareal, "Tesla's construction workers at Texas gigafactory allege labor violations," *The Guardian,* November 15, 2022. https://www.theguardian.com/technology/2022/nov/14/tesla-texas-construction-workers-gigafactory-lawsuit-labor-violations

[40] Tom Libby, "Surprising Trends in the US Electric Vehicle Market," S&P Global Mobility, June 18, 2024. spglobal.com/mobility/en/research-analysis/surprising-trends-us-electric-vehicle-market.html

[41] "How many electric vehicle charging stations are there in the US?" *USA Facts*, last updated March 28, 2023. https://usafacts.org/articles/how-many-electric-vehicle-charging-stations-are-there-in-the-us/

[42] Robert Rapier, "Environmental Implications Of Lead-Acid And Lithium-Ion Batteries," *Forbes*, January 19, 2020. https://www.forbes.com/sites/rrapier/2020/01/19/environmental-implications-of-lead-acid-and-lithium-ion-batteries/?sh=15af1ec27bf5

[43] Nathan Niese et al, "The Case for a Circular Economy in Electric Vehicle Batteries," *BCG*, September 14, 2020. https://www.bcg.com/publications/2020/case-for-circular-economy-in-electric-vehicle-batteries

[44] "Eilmer The Flying Monk," Athelstan Museum, accessed April 4 2023, https://www.athelstanmuseum.org.uk/malmesbury-history/people/eilmer-the-flying-monk/

[45] RG Grant, *Flight: 100 Years of Aviation.* (London, DK Publishing, October 2002)

[46] Bernd Lukasch, "The Aerodynamics of Lilienthal," *Otto Lilienthal Museum,* accessed March 2024. https://www.lilienthal-museum.de/olma/e34.htm

[47] TD Crouch et al., "History of Flight," *Encyclopaedia Britannica*, March 31, 2023, https://www.britannica.com/technology/history-of-flight

[48] "Wing warping," *Smithsonian Education,* accessed March 2024. https://smithsonianeducation.org/educators/lesson_plans/wright/wing_warping.html

[49] David Kindy, "This Odd Early Flying Machine Made History but Didn't Have the Right Stuff,". *Smithsonian Magazine,* May 5, 2021 https://www.smithsonianmag.com/smithsonian-institution/odd-early-flying-machine-made-history-didnt-have-right-stuff-180977658/

[50] Crouch et al., "History of Flight"

[51] Grant, *Flight*

[52] The Editors of Encyclopaedia Britannica, "Concorde," *Encyclopaedia Britannica*, last updated March 22, 2024. https://www.britannica.com/technology/Concorde

[53] "The Evolution of the Boeing 747 Fleet: 1966–2023," *IBA*, January 31, 2023. https://www.iba.aero/insight/evolution-of-the-boeing-747-fleet

[54] "Number of flights performed by the global airline industry from 2004 to 2021, with forecasts until 2023," Statistica, December 2022, https://www.statista.com/statistics/564769/airline-industry-number-of-flights/

[55] Jaap Bouwer et al., "Taking stock of the pandemic's impact on global aviation," McKinsey & Co, March 31, 2022, https://www.mckinsey.com/industries/travel-logistics-and-infrastructure/our-insights/taking-stock-of-the-pandemics-impact-on-global-aviation; Jaap Bouwer et al., "Back to the future? Airline sector poised for change post-COVID-19," McKinsey & Co, April 2, 2021. https://www.mckinsey.com/industries/travel-logistics-and-infrastructure/our-insights/back-to-the-future-airline-sector-poised-for-change-post-covid-19

[56] "Industry Statistics," *IATA*, December 2023. https://www.iata.org/en/iata-repository/pressroom/fact-sheets/industry-statistics

[57] Over 33,000 new planes valued over US$5 trillion for the next 20 years," Airbus press release, 11 July 2016, https://www.airbus.com/en/newsroom/press-releases/2016-07-over-33000-new-planes-valued-over-us5-trillion-for-the-next-20

[58] "Post-COVID-19 Forecasts Scenarios," International Civil Aviation Organization, accessed April 5, 2023, https://www.icao.int/sustainability/Pages/Post-Covid-Forecasts-Scenarios.aspx; Brian Prentice et al, "Global fleet and MRO market forecast 2022-2023," Oliver Wyman. https://www.oliverwyman.com/content/dam/oliver-wyman/v2/publications/2022/feb/MRO-2022-Master-file_v5.pdf

[59] "Carbon offsetting for international aviation," International Air Transport Association (IATA), June 2020, https://www.iata.org/contentassets/fb745460050c48089597a3ef1b9fe7a8/paper-offsetting-for-aviation.pdf

[60] Mishal Ahmad et al., "Opportunities for industry leaders as new travelers take to the skies," McKinsey & Co, April 5, 2022, https://www.mckinsey.com/industries/travel-logistics-and-infrastructure/our-insights/opportunities-for-industry-leaders-as-new-travelers-take-to-the-skies

[61] Derek Costanza and Brian Prentice, "Aviation Growth is Outpacing Labor Capacity," *Velocity Travel, Transport and Logistics 2017*, Oliver Wyman LLC, accessed April 2023, https://www.oliverwyman.com/our-expertise/insights/2017/sep/oliver-wyman-transport-and-logistics-2017/operations/aviation-growth-is-outpacing-labor-capacity.html

[62] "Carbon offsetting for international aviation," IATA, accessed April 2024. https://www.iata.org/contentassets/fb745460050c48089597a3ef1b9fe7a8/paper-offsetting-for-aviation.pdf

[63] "Boeing Doubles Sustainable Aviation Fuel Purchase for Commercial Operations, Buying 5.6 Million Gallons for 2023," *Boeing*, February 15, 2023. https://investors.boeing.com/investors/news/press-release-details/2023/Boeing-Doubles-Sustainable-Aviation-Fuel-Purchase-for-Commercial-Operations-Buying-5.6-Million-Gallons-for-2023/default.aspx#:~:text=SAF%20reduces%20CO2%20emissions,next%2020%20to%2030%20years.

⁶⁴ Tim Bowler, "Carbon fibre planes: Lighter and stronger by design," *BBC News,* January 28, 2014; https://www.bbc.co.uk/news/business-25833264 "The BAMCO consortium: to develop biosourced composites using bamboo fibres," *Expleo.* November 19, 2018. https://expleo.com/global/en/case-studies/bamco-bamboo-long-fibre-reinforced-biobased-matrix-composites/

⁶⁵ "Carbon offsetting for international aviation," IATA, accessed March 2024. https://www.iata.org/contentassets/fb745460050c48089597a3ef1b9fe7a8/paper-offsetting-for-aviation.pdf

⁶⁶ "Carbon offsetting," IATA.

⁶⁷ "What's new with Boeing sustainability," *Boeing*, accessed March 2024.) https://www.boeing.com/sustainability#new

⁶⁸ "Sustainable Aviation Fuel Fact Sheet," *Boeing*, April 2023. https://www.boeing.com/content/dam/boeing/boeingdotcom/principles/esg/SAF-fact-sheet.pdf

⁶⁹ "Boeing Isn't Grounding Airline Reputations," Eleanor Hawkins, *Axios,* May 24, 2024, https://www.axios.com/2024/05/24/boeing-airline-reputation-harris-poll

⁷⁰ "Flight Efficiency and Sustainability," *Boeing*, accessed March 2024. https://services.boeing.com/flight-operations/flight-efficiency-sustainability

⁷¹ "Airbus – net-zero carbon emissions by 2050," *ERT*, accessed April 2024. https://industry4climate.eu/casestudy/airbus/

⁷² "ZEROe "Towards the world's first hydrogen-powered commercial aircraft," Airbus, accessed March 2024. https://www.airbus.com/en/innovation/low-carbon-aviation/hydrogen/zeroe#:~:text=Airbus'%20ambition%20is%20to%20bring,produce%20and%20supply%20the%20hydrogen; Jasmine Jessen, "Airbus: Hydrogen Aircraft Could Be Delayed by a Decade," *Sustainability* magazine, February 20, 2025, https://sustainabilitymag.com/articles/airbus-delays-plans-for-commercial-hydrogen-aircraft.

⁷³ "Decarbonisation - Towards low-carbon air travel for future generations," *Airbus*, accessed March 2024. https://www.airbus.com/en/sustainability/respecting-the-planet/decarbonisation

⁷⁴ Airbus, "Sustainable aviation fuels," accessed March 2024. https://www.airbus.com/en/sustainability/respecting-the-planet/decarbonisation/sustainable-aviation-fuels#:~:text=All%20Airbus%20aircraft%20are%20capable,technical%20pathways%20for%20producing%20SAF.

[75] "Responsible supply chain," *Airbus*, accessed March 2024. https://www.air-bus.com/en/sustainability/respecting-the-planet/responsible-supply-chain#:~:text=Approximately%208%2C000%20direct%20and%2018%2C000,communities%20in%20which%20they%20operate.

[76] "Global carbon standard for airports passes the 500 milestone," *Airport Carbon Accreditation,* May 30, 2023. https://www.airportcarbonaccreditation.org/global-carbon-standard-for-airports-passes-the-500-milestone/#:~:text=Olivier%20Jankovec%2C%20Director%20General%20of,the%20Airport%20Carbon%20Accreditation%20framework.

[77] "Ocean shipping and shipbuilding," Organisation for Economic Co-operation and Development, accessed April 5, 2023, www.oecd.org/ocean/topics/ocean-shipping/

[78] Naya Olmer et al., "Greenhouse Gas Emissions From Global Shipping, 2013-2015, "International Council on Clean Transportation, October 2017, https://theicct.org/sites/default/files/publications/Global-shipping-GHG-emissions-2013-2015_ICCT-Report_17102017_vF.pdf

[79] "Revised GHG reduction strategy for global shipping adopted," *IMO*, July 7, 2023. https://www.imo.org/en/MediaCentre/PressBriefings/pages/Revised-GHG-reduction-strategy-for-global-shipping-adopted-.aspx

[80] "Clean green marine: the race to achieve zero-emission shipping by 2050," Expleo Group, accessed April 2023, https://expleo.com/global/en/insights/whitepapers/clean-green-marine-the-race-to-achieve-zero-emission-shipping-by-2050/

[81] "These emerging economies are poised to lead shipping's net-zero transition," *World Economic Forum*, August 18, 2022. https://www.weforum.org/agenda/2022/08/maritime-shipping-decarbonization-emerging-economies/

[82] "Clean Maritime Plan," Department for Transport, July 2019. https://assets.publishing.service.gov.uk/media/5d24a96fe5274a2f9d175693/clean-maritime-plan.pdf

[83] AP Møller – Mærsk, accessed April 2023, https://www.maersk.com/news/articles/2023/05/17/e-book-reduce-your-carbon-footprint-in-fmcg-logistics

[84] "IMO 2020 – cutting sulphur oxide emissions," IMO, accessed April 2023 https://www.imo.org/en/MediaCentre/HotTopics/Pages/Sulphur-2020.aspx

[85] https://datamillnorth.org, accessed April 2023.

86 https://amsterdamsmartcity.com, accessed April 2023.

87 https://www.spatial.nsw.gov.au/digital_twin, accessed April 2023.

88 Phil Goldstein, "How Los Angeles Plans to Become a Smarter City by the 2028 Olympics," *State Tech Magazine*, July 29, 2021, https://statetechmagazine.com/article/2021/07/how-los-angeles-plans-become-smarter-city-2028-olympics

89 Pascual Berrone and Joan Enric Ricart "IESE Cities in Motion Index 2022," IESE Business School, accessed April 2023 https://media.iese.edu/research/pdfs/ST-0633-E.pdf.

5. Why Sustainability Became a 21st-Century Force for Good

1 Curt Mueller et al, "Operations-driven sustainability", McKinsey & Company, 10th August 2020, www.mckinsey.com/capabilities/operations/our-insights/operations-driven-sustainability

2 Jennifer Tonti, Americans Want to See Greater Transparency on ESG Issues and Support Federal Requirements for Increasing Disclosure," Just Capital, February 2022. https://com-justcapital-web-v2.s3.amazonaws.com/pdf/JUSTCapital_CorporateDisclosureStandardsSurveyReport_SSRS_Ceres_PublicCitizen_Feb2022.pdf

3 Witold Henisz et al., "Five ways that ESG creates value," *McKinsey Quarterly*, November 2019. https://www.mckinsey.com/~/media/McKinsey/Business%20Functions/Strategy%20and%20Corporate%20Finance/Our%20Insights/Five%20ways%20that%20ESG%20creates%20value/Five-ways-that-ESG-creates-value.ashx

4 Debra Brown and David Brown, *ESG Matters: How to Save the Planet, Empower People, and Outperform the Competition* (Ethos Collective, August 24, 2021)

5 SASB Standards Materiality Finder. https://sasb.org/standards/materiality-finder

6 Respect International "G4 Sustainability Reporting Guidelines," accessed March 2024, https://respect.international/wp-content/uploads/2017/10/G4-Sustainability-Reporting-Guidelines-Implementation-Manual-GRI-2013.pdf

[7] Mozaffar Khan et al, "Corporate Sustainability: First Evidence on Materiality" *The Accounting Review*, Vol. 91, No. 6, pp. 1697-1724., November 9, 2016, https://ssrn.com/abstract=2575912 or http://dx.doi.org/10.2139/ssrn.2575912

[8] Gordon L. Clark et al, "From the stockholder to the stakeholder: how sustainability can drive financial performance," Arabesque Partners, last updated March 2015, https://arabesque.com/research/From_the_stockholder_to_the_stakeholder_web.pdf

[9] Alex Edmans, "The Link Between Job Satisfaction and Firm Value, With Implications for Corporate Social Responsibility," *SSRN*, last revised December 19, 2013. https://papers.ssrn.com/sol3/papers.cfm?abstract_id=2054066

[10] "Our Story," The Body Shop International Limited, accessed April 26, 2023, https://www.thebodyshop.com/en-gb/about-us/our-story/a/a00002.

[11] "Pampers: The Birth of P&G's First 10-Billion-Dollar Brand," June 27, 2012. https://www.pg.co.uk/blogs/pampers-birth-pgs-first-10-billion-dollar-brand/

[12] Charlotte Edmond, "Disposable nappies are one of the biggest contributors to plastic waste – but how green are the alternatives?" World Economic Forum, August 23, 2023. https://www.weforum.org/agenda/2023/08/disposable-nappies-landfill-plastic-circular-economy/#:~:text=Plastic%20fibres%2C%20likely%20from%20nappies,harm%20children%20and%20the%20environment.

[13] Matthew Taylor, "'Total monster': fatberg blocks London sewage system," *The Guardian*, September 12, 2017. https://www.theguardian.com/environment/2017/sep/12/total-monster-concrete-fatberg-blocks-london-sewage-system

[14] "Pampers Sustainability – Love less waste," *Pampers*, August 1, 2022. https://www.pampers.co.uk/safety-and-commitment/pampers-commitments/article/pampers-sustainability-love-less-waste

[15] "Reducing Our Environmental Impact," *Pampers*, last updated May 7, 2020. https://www.in.pampers.com/about-pampers/better-for-baby/article/reducing-our-environmental-impact

[16] Dalrymple, *The Anarchy. The Relentless Rise of the East India Company*, (New York, NY: Bloomsbury Publishing, 2019)

[17] A Cadbury, *Report of the Committee on the Financial Aspects of Corporate Governance* (London: Gee, 1992), accessed April 26, 2023. https://www.frc.org.uk/getattachment/9c19ea6f-bcc7-434c-b481-f2e29c1c271a/The-Financial-Aspects-of-Corporate-Governance-(the-Cadbury-Code).pdf

18 "Taking CFCs Out of Aerosols. How Sam Johnson Led SC Johnson to Environmental Activism," SC Johnson, accessed April 26 2023, https://www.scjohnson.com/en/about-us/the-johnson-family/sam-johnson/taking-cfcs-out-of-aerosols-how-sam-johnson-led-sc-johnson-to-environmental-activism

6. The Sustainability Maturity Scale

1 Christian Heller, "Value to Society – a Balanced Approach to Measuring Business Impact," The Global Goals Yearbook 2019, accessed April 26 2023, https://www.basf.com/global/documents/en/sustainability/we-drive-sustainable-solutions/quantifying-sustainability/we-create-value/Global%20Goals%20Yearbook_2019_BASF_2019-08-05%20(1).pdf

2 Mike Scott, "How a one-man scrap metal recycler became the world's most sustainable corporation," Corporate Knights, January 18, 2023. https://www.corporateknights.com/rankings/global-100-rankings/2023-global-100-rankings/top-company-profile-schnitzer-steel/

3 Sphera Editorial Team, "Challenges and Opportunities in Growing a Green Economy," Sphera, January 25, 2023, https://sphera.com/report/green-economy-outlook-attitudes-toward-sustainability-among-consumers-and-operations-managers/

4 European Green Digital Coalition, last updated January 24, 2023, https://digital-strategy.ec.europa.eu/en/policies/european-green-digital-coalition

5 Corporate Sustainability Reporting Directive, accessed March 2024. https://finance.ec.europa.eu/capital-markets-union-and-financial-markets/company-reporting-and-auditing/company-reporting/corporate-sustainability-reporting_en

6 "CSRD & Supply Chain Sustainability: What You Need to Know," *Supply Shift*, February 24, 2023. https://www.supplyshift.net/blog/csrd-supply-chain-sustainability-what-you-need-to-know/

7 Michael Greenstone, Christian Leuz, and Patricia Breuer, "Mandatory Disclosure Would Reveal Corporate Carbon Damages," *Science*, 24 August 2023, Vol. 381, Issue 6660, pages 837-840.

7. Active Stakeholder in Your Ecosystem

[1] James F Moore, "Predators and Prey: A New Ecology of Competition", *Harvard Business Review,* (May-June 1993) https://hbr.org/1993/05/predators-and-prey-a-new-ecology-of-competition

[2] "Why business must harness the power of purpose," Ernst & Young, December 15, 2020, https://www.ey.com/en_uk/purpose/why-business-must-harness-the-power-of-purpose

[3] "Companies that work together for the common good? That's coopetition!," Saint Gobain, October 12, 2020, https://www.saint-gobain.com/en/magazine/stories/companies-work-together-common-good-thats-coopetition

[4] Michael Tran, "How Much Does Employee Retention Impact Customer Satisfaction?," Medallia, February 16, 2016, https://www.medallia.com/blog/how-much-does-employee-retention-impact-customer-satisfaction/

[5] Megan Carnegie, "Are chief sustainability officer jobs women's fast-track to the C-suite?" March 8, 2024. https://www.bbc.com/worklife/article/20240307-women-chief-sustainability-officer-jobs

[6] Einar H Dyvik, "The 100 largest companies in the world ranked by revenue in 2023," *Statista,* February 13, 2024. https://www.statista.com/statistics/263265/top-companies-in-the-world-by-revenue/#:~:text=Top%20companies%20in%20the%20world%20by%20revenue%202023&text=With%20nearly%20640%20billion%20U.S.,million%20all%20over%20the%20world.

[7] Margot Meeks and Rachel JC Chen, "Can Walmart Integrate Values with Value?: From Sustainability to Sustainable Business" *Retail, Hospitality, and Tourism Management Publications and Other Works*, 2011, accessed April 26, 2023, https://trace.tennessee.edu/utk_retapubs/4

[8] Humes, Edward, *Force of Nature: The Unlikely Story of Wal-Mart's Green Revolution,* (New York, NY: Harper Business, May 2011)

[9] The Sustainability Consortium. Accessed March 2024. https://sustainabilityconsortium.org

[10] Kathleen McLaughlin, "Accelerating Climate Action: Project Gigaton™ Marks Key Milestone," *Walmart*, April 6, 2022.

https://corporate.walmart.com/news/2022/04/06/accelerating-climate-action-project-gigaton-marks-key-milestone

[11] Walmart Sustainability, accessed March 2024. https://corporate.walmart.com/purpose/sustainability

[12] "Environmental, Social and Governance Summary Report FY2022," Walmart, accessed April 26, 2023, https://corporate.walmart.com/content/dam/corporate/documents/purpose/environmental-social-and-governance-report-archive/walmart-fy2022-esg-summary.pdf

[13] Daniel Goleman, "Handprints, Not Footprints," *Time*, March 12, 2012. https://content.time.com/time/subscriber/article/0,33009,2108015,00.html

[14] Navi Radjou, "Beyond Sustainability: The Regenerative Business," *Forbes.com*, October 24, 2020, https://www.forbes.com/sites/naviradjou/2020/10/24/beyond-sustainability-the-regenerative-business/?sh=688bd481ab35

[15] "Plants Use An Internet Made of Fungus," accessed April 26, 2023, https://ed.ted.com/best_of_web/4uORORJx

[16] Andrew Edgecliffe-Johnson, "How Walmart convinced critics it can sell more stuff and save the world," *Financial Times,* October 13, 2022 https://www.ft.com/content/0975d1e3-d95f-4b77-8d58-5d95a751f31a

[17] "Wegmans: A family company since 1916.", accessed March 2024. https://www.wegmans.com/wp-content/uploads/wegmans-company-history-061522.pdf

[18] Wegmans Awards Record Number of Employee Scholarships," Wegmans, accessed April 26, 2023, https://www.wegmans.com/news-media/articles/wegmans-awards-record-number-of-employee-scholarships

[19] Sustainable Living Plan 2010 to 2020: Summary of 10 Year's Progress. https://www.unilever.com/files/92ui5egz/production/16cb778e4d31b81509dc5937001559f1f5c863ab.pdf

[20] Gil Press, "The Ultimate Entrepreneur: Jim Goodnight," *Forbes.com*, September 27, 2017, https://www.forbes.com/sites/gilpress/2017/09/27/the-ultimate-entrepreneur-jim-goodnight-sas/

21 "SAS CEO Jim Goodnight on SAS' Great Employee Benefits," accessed April 26, 2023. https://www.youtube.com/watch?v=T5O3L6UdIGw.

22 "Sustainability," *United Nations Academic Impact*, accessed April 2024. https://www.un.org/en/academic-impact/sustainability

23 "Social and Environmental Values Increasingly Drive Consumers' Choices, According to New Research," Globe Scan press release, June 23, 2021, https://globescan.com/2021/06/23/social-environmental-values-increasingly-drive-consumers-choices/

24 Dow Jones Sustainability World Index, accessed March 2024. https://www.spglobal.com/spdji/en/indices/esg/dow-jones-sustainability-world-index/#overview

25 Geoffrey Guest, "LCA INC. a brief history of LCA in the corporate world," *LinkedIn Pulse*, July 25, 2015, https://www.linkedin.com/pulse/lca-inc-geoffrey-guest/

26 Ian Williams and Alice Brock, "Ranked: the environmental impact of five different soft drink containers," *The Conversation,* November 17, 2020, https://theconversation.com/ranked-the-environmental-impact-of-five-different-soft-drink-containers-149642

27 "Ambition 2039," *Mercedes-Benz Group*, accessed March 2024. https://group.mercedes-benz.com/responsibility/sustainability/climate-environment/ambition-2039-our-path-to-co2-neutrality.html#:~:text=Net%20carbon%2Dneutral%20production%20since%202022&text=The%20ambition%20for%20all%20Mercedes,are%20to%20go%20into%20operation.

28 "Quantifying supply chain decarbonization as a pillar for becoming a net-zero company," *Sphera*, accessed April 2024. https://sphera.com/wp-content/uploads/2021/07/CS-Daimler-Mercedes-Benz.pdf

29 "Single-use plastic bags and their alternatives - Recommendations from Life Cycle Assessments," United Nations Environment Programme (2020), accessed April 26, 2023, https://www.lifecycleinitiative.org/wp-content/uploads/2021/03/SUPP-plastic-bags-meta-study-8.3.21.pdf

8. Supply Chains for Good

[1] Global Positioning Satellites, Enterprise Resource Planning, Radio Frequency Identification, Lifecycle Assessment, Artificial Intelligence, Machine Learning and Internet-of-Things

[2] Talmage Wagstaff, "Minimizing the Effect Of Downtime," *Manufacturing.net*, June 30, 2021. https://www.manufacturing.net/operations/blog/21533375/minimizing-the-effect-of-downtime

[3] Anne-Titia Bové and Steven Swartz, "Starting at the source: Sustainability in supply chains." *McKinsey Sustainability*, November 11, 2016. https://www.mckinsey.com/capabilities/sustainability/our-insights/starting-at-the-source-sustainability-in-supply-chains

[4] The Sustainability Insight System (THESIS), accessed March 2024. https://sustainabilityconsortium.org/thesis

[5] "Signify Automates, Standardizes and Innovates Supply Chain Risk Management," accessed March 2024; https://sphera.com/wp-content/uploads/2023/05/03162022-Signify-Case-Study_v2.pdf

[6] Jeremy Kingsley "The Business Costs of Supply Chain Disruption." *The Economist,* February 25, 2021. https://impact.economist.com/perspectives/sustainability/business-costs-supply-chain-disruption-1

[7] Susan Lund et al. "Risk, resilience, and rebalancing in global value chains." *McKinsey &* Company, August 6, 2020. https://www.mckinsey.com/capabilities/operations/our-insights/risk-resilience-and-rebalancing-in-global-value-chains

An Action Plan for Positive Change

[1] "Gartner Survey Finds 87% of Business Leaders Expect to Increase Sustainability Investment Over the Next Two Years." Gartner newsroom. November 14, 2022. https://www.gartner.com/en/newsroom/press-releases/2022-11-14-gartner-survey-finds-87-percent-of-business-leaders-expect-to-increase-sustainability-investment-over-the-next-two-year

INDEX

Index

Index

ABOUT THE AUTHOR

As Sphera's founding CEO and president, Paul Marushka is responsible for providing overall strategic leadership for the company in developing, directing and implementing go-to-market, service, product, and operational plans.

Paul has grown businesses by bringing innovative solutions to market, leveraging software, analytics, and technology services. Prior to the founding of Sphera, Paul served as president of Marsh ClearSight, a business unit of Marsh & McLennan, which is a leading provider of software, services, and analytics for enterprise risk management, safety and compliance management, and claims administration. Paul also has held executive positions at software and data companies such as Fair Isaac Corp. (FICO) and CCC Information Services.

During his career, Paul has developed and launched a variety of software and analytics products recognized by the Gartner Group for their impact on the industry. He has also authored numerous articles in publications, including the *United Nations Global Yearbook*, on the use of analytics and technology in decision-making. He was awarded an EY Entrepreneur of the Year 2021 Midwest Award for his dedication and leadership.

Paul has a J.D. from Northwestern University Pritzker School of Law, an MBA from the University of Chicago Booth School of Business, and an A.B. from the University of Chicago.